BIRD SENSES

BIRD SENSES

How and What Birds See, Hear, Smell, Taste, and Feel

GRAHAM R. MARTIN

PELAGIC PUBLISHING

Published by Pelagic Publishing
PO Box 874
Exeter
EX3 9BR
UK

www.pelagicpublishing.com

Bird Senses: How and What Birds See, Hear, Smell, Taste, and Feel

ISBN 978-1-78427-216-6   *Paperback*
ISBN 978-1-78427-217-3   *ePub*
ISBN 978-1-78427-218-0   *PDF*

Front cover: Great Grey owl (*Strix nebulosa*) taking off from the
ground, Finland, April. © Danny Green/naturepl.com
Rear cover: White-tailed Eagle, Creative Commons, Sony

Printed and bound in India by Replika Press Pvt. Ltd.

MIX
Paper from
responsible sources
FSC® C016779
FSC
www.fsc.org

For Amber, Bryher, Josie, and Ted

Contents

Preface   viii

A note on bird taxonomy and bird names   x

Acknowledgements   x

1.   Senses and sensory ecology   1
2.   Investigating senses   11
3.   Vision in birds: the basics   40
4.   Bird eyes: variations and consequences   61
5.   Visual fields   85
6.   Beyond vision: hearing and smell   115
7.   The intimate senses: touch and taste   148
8.   Sensing the earth's magnetic field   166
9.   Birds in the dark   172
10. Other birds of the night: parrots to passerines   200
11. Birds underwater   213
12. A sideways look through birds' eyes   240

Appendix: Visual acuity in birds   259
Further reading   263
Index   264

Preface

This book's aim is to take the reader deep into the world of birds. We may think we know the world of birds through our own senses. But we are just one species, and we are not birds! To get into the world of birds we must go beyond the habitat or ecological points of view. It is necessary also to take a sensory perspective. It is a perspective that leads to an understanding of the different kinds of information that birds have available to them as they live out their lives beside us.

This is a 'through birds' eyes' approach to ornithology – but as we shall see there is a lot more to a bird's world than the information that it receives from its visual system. Other sensory information is constantly in play and interacting to provide each species with a unique suite of information that guides its daily activities. As the book dives deeper into the senses of birds and their diversity across species, we become aware that our world, the human world, is just one way of extracting information from the environment that surrounds us. What we consider to be reality is but one of many worlds in which species live as they rub alongside each other.

As in all aspects of bird physiology, behaviour, and ecology, diversity is the name of the game. The senses of each bird, and the information that they provide, have been uniquely tuned through natural selection to solve the challenges of different environments, the challenges of different foraging techniques, and the exploitation of different resources. Of course there are commonalities across species, but wherever we look there are differences.

While we may ponder the intriguing diversity of birds' bills and describe how they are tools for exploiting different resources in optimal ways, it is easy to overlook that those tools need guidance. In each bird species, the senses have become tuned to provide optimal information that guides behaviour, day in, day out.

The book first sets out some basic information about each of the main senses. However, from the start, questions are asked about how senses can and do differ between species. This should help the reader to get a grasp of how to compare and investigate different senses in different species, and to form questions about what information different species might have available from moment to moment to guide their behaviour. What is known about each of the main sensory systems of birds is uneven; some species or sensory systems are known about in detail, others less so. Of course, with almost 11,000 species of birds to investigate, it is not surprising that very little is known about many of them. But that is true of all aspects of their ecology, behaviour, and physiology. We have to piece together general truths from patchy information.

The later chapters bring together what is known across different species, to offer explanations of how birds cope with particular environmental challenges. I have focused particularly on activity at night and activity underwater, both of which provide very different challenges to what we may think of as birds' normal daytime terrestrial environments. In dwelling on these topics, it is perhaps surprising to discover how often birds are prepared to act guided only by sketchy information, as they operate at the limits of their sensory systems.

Finally, there is discussion of the many sensory challenges that birds face because of the obstacles that humans have contrived to put in their way, from power lines to windowpanes. A sensory ecology approach does suggest some answers to those challenges, and so the book ends on an optimistic note. It shows how understanding the senses of birds, and the information that they provide, can suggest solutions to the problems that humans have presented them with.

My aim has been to make technical information understandable and the book readable for the keen birdwatcher and naturalist, as well as the more specialist reader. I have brought together a wealth of diverse information, but the book is not laden with references. Some of the more difficult topics are explained in boxes that can be referred to separately from the rest of the text. However, the reader can be assured that all statements are well supported by published material. There are suggestions for further reading which can be followed up should the reader be intrigued to dive deeper into the sensory worlds of birds.

Graham Martin
Emeritus Professor of Avian Sensory Science
University of Birmingham, UK
October 2019

A note on bird taxonomy and bird names

The taxonomy and naming of birds is a matter of much debate. There are rival taxonomies and there can be heated debates. Furthermore, taxonomic revisions are frequent, especially now that classification based on genetic data is firmly established. The taxonomy, scientific and English names used in this book follow those of the International Ornithology Congress World Bird List (www.worldbirdnames. org). This is a freely open resource, with a lot of useful information, lists and spreadsheets, that can be downloaded. It is frequently updated. Names used in this book are based on version 9.2 (doi: 10.14344/IOC.ML.9.2) – but by the time you read this book it is likely that a newer version will be available from the IOC.

Acknowledgements

I have studied the sensory world of birds for a lifetime and have worked alongside a wonderful range of keen and enthusiastic people. They have helped me to investigate a fantastic range of birds and their senses, and in many different locations. I thank them all for their support, and the exchange of ideas and enthusiasm. The names of them will have appeared as joint authors of papers or in the acknowledgement footnotes of papers. They have all contributed to this book in different ways, and I acknowledge their help and encouragement. Various colleagues have read parts of this book, but one person who has read it all and sorted out overcomplicated language and many grammatical errors is Judith Burl, and I especially thank her for helping to bring the book to completion. I must also acknowledge the very skilful and helpful copy-editing of Hugh Brazier. With an eagle's eye he caught my errors and checked many details. He gave the manuscript a deep clean and final polish. Finally, I thank people who over the years have attended talks that I have given to both specialist and lay audiences. They have asked intriguing questions and challenged me to explain more. In so doing they convinced me that there could be a readership for a book on this topic.

Chapter 1

Senses and sensory ecology

A Peregrine leaves its lookout in pursuit of a dove that it detected from more than a kilometre away. In under two minutes the dove is held firmly in the falcon's talons. An Oilbird flies to its nest ledge inside a cave that is so deep that no light enters, darkness is total. In moonlight a Red Knot probes its bill through the squelchy surface of estuary mud, and without even lifting its bill detects and ingests a buried worm. A Great Tit searches through a wood and locates an infestation of caterpillars on a particular tree. The presence of the caterpillars is detected well before the Great Tit sees them.

All of these are brilliant examples of bird senses in action. Each describes an instance of a bird using particular information to control one of its key behaviours. Included in these examples are the detection of prey at a great distance, precise seizure of a target by the feet or bill, location of a foraging site, mobility in the dark. Each bird's survival requires that these actions are executed many times during its lifetime, and each action must be executed with high accuracy, both in time and in space.

These examples were chosen because each depends on information gathered primarily by a different sense. Vision in the case of the Peregrine, hearing in the Oilbird, touch and taste in the Knot, and olfaction (the sense of smell) in the Great Tit. While information from one sense underpins each of these particular behaviours, these birds must also rely on information from a suite of different senses in order to conduct themselves safely during their everyday lives (Figure 1.1).

In all instances this different information is integrated and interpreted by the birds' brains in a seamless fashion. We may focus our attention on the pursuit flight of the Peregrine, and wonder how it achieves such an impressive performance, but in reality the bird is moving rapidly from the execution of one task to another. No sooner has the prey been spotted than it is time to line up for its seizure, and the prey is grabbed with impeccable timing. We can even think of these birds as simultaneous multitaskers, for example looking out for predators or a competitor while searching for food, and each different task requires different information.

Equally important is the range of environmental conditions in which these four birds carry out their key tasks – from total darkness to bright sunlight, from open habitats to the structural complexity of a forest. We could also choose examples of birds foraging in the open airspace, or finding food hidden on hard and soft

FIGURE 1.1 Clockwise from top left: a Peregrine *Falco peregrinus*, an Oilbird *Steatornis caripensis*, a Red Knot *Calidris canutus*, and a Great Tit *Parus major*. Each species exploits information from a different sense to guide some its key behaviours. (Photograph of Great Tit by Francis C. Franklin [CC-BY-SA-3.0].)

surfaces, or beneath surfaces, or through the surface of water, or underwater at both shallow and deep depths, so deep that daylight hardly penetrates. Each habitat type presents different sensory and informational challenges which the birds must meet in order to live out their daily lives.

Every bird species is special. By definition every species is unique, and much of what attracts our attention to particular birds is the fact that they are able to execute tasks that others cannot. It is this diversity of behaviour, as much as diversity of appearance, that makes watching and studying birds so rewarding.

At one site, on one day, even at the same moment, we can see birds of different species acting in very different ways. Each is a specialist, and it is the parade of these specialisms that is behind much of the allure of watching birds. But while we can readily understand how obvious structural adaptations, particularly of their wings, bills, and feet, can allow different species to execute particular actions, it is easy to overlook the fact that those wings, bills, and feet must be guided accurately.

There is no value in having a tool if it cannot be adequately controlled. A hammer that cannot be brought down accurately to hit a nail is useless. The hammer needs precision guidance in both time and space. In the same way, a long bill for grabbing a fish, or a short bill for seizing a seed, or a foot for grabbing another bird out of the air, are useless if they cannot be controlled and brought exactly to the target with precise timing. Each bill, foot, and wing requires specialised information to guide it to the right place and to get there at the right time. Actions need information, and different actions need different information.

Unlocking the information

The information that each bird species uses to guide its behaviours is a kind of secret. It is information that only the bird itself has direct access to. But for us to understand birds properly those secrets need to be unlocked in some way. The information that birds employ is not readily available to us as we look on. We need some tricks to help us get more than a hint of the information that a bird might be acting upon at any moment.

As humans, we are trapped in our own world, with its own secrets. Our eyes do not allow us to see what the Peregrine sees as it bears down upon a pigeon. We cannot feel what is at the Red Knot's bill tip, we cannot smell the odours given off by leaves as they are devoured by caterpillars, and we cannot interpret the echoes of an Oilbird's call as it bounces from the wall of a light-tight cave.

Even if we could look from the same perspective as the hunting Peregrine, we would not see and detect the world as it does. A view that is based upon the same vantage point as the bird's cannot tell us what the Peregrine actually sees. Our eyes are not the eyes of a Peregrine, our ears are not those of an Oilbird, our fingertips are not able to detect what the Red Knot's bill feels, and our tongues cannot taste what the Knot's tongue tastes. Because of differences in the senses of different species, the information they provide of the same scene is different. In effect each species lives in a different secret world. Species may share the same environment, but the worlds that they inhabit are different.

Our inability to experience the different worlds of birds is not because our senses are relatively impoverished; it is not because birds have 'super-senses'. There are certainly examples of bird species that can out-perform us in particular ways, but equally our senses often out-perform those of particular birds. The problem is that our own senses are specialised for the conduct of particular actions just as much as the senses of animals are specialised for the control of their particular actions.

Despite the popular view that humans are such an important species, it is clear that our eyes are not all-seeing, and our ears are not the detectors of every sound. All senses are selective of the information that they retrieve from the environment. One advantage that humans have over the birds, however, is that we can get valuable insights into their worlds. We can unlock some of their secrets through the application of science. In doing so we not only gain insights into the worlds of birds, but we also come to understand our own world from a fresh perspective.

Sensory ecology

Sensory ecology is the branch of science that enables us to understand how information controls animals' interactions with their environments. In essence sensory ecology allows us to enter and understand the worlds of other species. When we apply the ideas of sensory ecology to birds, we move close to appreciating the world 'through birds' eyes'. We also come to understand the proper meaning of a 'bird's-eye view'.

A bird's-eye view is a much-used metaphor. It pops up everywhere, from highbrow literature, to journalism, to advertising. It is shorthand for appreciating the world from a fresh perspective, but often it is used to imply that we can actually see the world through the eyes of a bird. Sensory ecology casts doubt on that very notion. By going beyond just vision and eyes, a metaphorical bird's-eye view throws light upon all senses and on how information from them is integrated in the conduct of particular tasks. In so doing it helps us delve deep into the problem of understanding 'reality' from a human perspective (Figure 1.2).

Sensory ecology tells us about the ways in which different animals gather information from their environments, about the factors that ultimately limit what can be detected, and the ways in which information from different senses is brought together to underpin life in different habitats. Sensory ecology reveals to us that

FIGURE 1.2 A Golden Eagle *Aquila chrysaetos* stares into a camera lens. A White-tailed Eagle *Haliaeetus albicilla* flies over Paris carrying a camera as part of an advertising stunt to market a new camera. The advertisement carried the strapline, 'The 100% eagle-eye view'. These are two species credited with the most acute vision in all animals, but what do they actually see? What is an eagle's-eye view? Can it be captured in a single photographic image? (White-tailed Eagle from Sony Creative Commons; Golden Eagle by the author.)

there are very many 'birds'-eye views', many different sensory worlds, and that none is the same as ours. Our own sensory ecology is as specialised as those of Peregrines, Oilbirds, Knots, or Great Tits. We may wish to understand the world through their senses, but we can only ever experience one view of the world, our own. The best we can do as regards the worlds of other species is to know *about* them, we cannot ever experience and properly *know* their worlds.

Sensory ecology can provide data on the sensory performance of other species. It can place these data in a comparative framework and tell us how species differ one from another. It is often tempting to use the terms 'better' or 'worse' when comparing the sensory capacities of species but, on the whole, these terms are not useful. This is because the senses of each species are adapted for the conduct of different tasks in different environments. There may be optimal solutions to different sensory challenges, but they are no better or worse than each other. They are fit for the purposes that they evolved to meet.

Sensory ecology helps to determine why and how performance is limited by different environments and by the physiology of a species. Sensory ecology also indicates how information from one sense might be traded off against that provided by other senses for the optimal conduct of particular actions.

Modern technology can help in the tasks of characterising the sensory performance of different species, but it must not be thought capable of providing simple answers. Experiencing the world as a Peregrine sees it cannot be achieved by strapping a video camera to its back. Such footage makes thrilling viewing and gives insights into the life of a Peregrine, but it cannot give us the real Peregrine's-eye view. The photograph or film, viewed on a flat screen, still gives us only a view of the world that is two-dimensional and ultimately filtered by human eyes and brains. Understanding a bird's-eye view has to be achieved by piecing together many different sources of information about vision, not to mention about how visual information interacts with information from other senses.

Sensory ecology also shows us that we need to see the behaviours of birds in their proper context, the context of the actual challenges that birds face in their natural environments. For example, what is natural night-time really like? How does it change with habitat or with the annual cycle? We can also ask under what natural light conditions a Peregrine can best detect distant prey, and how changing light levels affect its abilities. Does our knowledge of these explain when and how it hunts? Similarly, we might ask questions about how soft and wet mud has to be for a Knot to be able to detect and identify a food item that is near the tip of its probing bill. We might also ask why some birds are particularly prone to flying into power lines. Is it due to limitations of their senses?

To answer all of these questions, and very many more, requires the senses of birds to be seen in both an ecological and an evolutionary context. In other words, knowledge of senses gained through careful laboratory-based studies and field observations needs to be placed in the context of the wider world in which animals evolved and exist today.

Key lessons from sensory ecology

Sensory ecology is a multidisciplinary field of endeavour. It pieces together information from many sources gained through different techniques and relies a lot on comparisons between species. Often it is able to answer some specific questions about the behaviour of birds, but general principles are rather few. However, there are two key themes which shall be returned to a number of times in this book. These themes are the role of multiple information sources in the control of particular behaviours, and the reliance upon minimal information for the control of many actions. Examples of these themes will be discussed throughout the book, but it is worth introducing them in general terms now.

Multiple information

Sensory ecology suggests that information from one sense can often be considered to predominate in the control of the behaviour of a particular species. We might even be tempted to assume that birds are 'all about vision'. However, information from other senses frequently comes into play, sometimes in unexpected ways. Information from different senses may often be traded off and used in complementary ways for the successful completion of a particular action.

This is particularly seen to be the case when birds are active at low light levels, but it also applies to situations when birds execute tasks at high light levels. For example, a high-speed video will reveal the great precision with which a hovering hummingbird inserts its bill into a flower's tube in search of nectar. Are the final manoeuvres guided by vision, by touch cues from the bill, or by taste, or is information from all three working in concert? Does the bird stop moving forward as soon as it tastes the nectar, or is it when the bill tip touches the nectary, or are visual cues from the flower entrance used?

As we consider the role of sensory information across the full diversity of birds, it becomes clear that many commonly observed behaviours involve not only sight but also hearing, taste, touch, smell, and even the detection of the earth's magnetic field, used in various combinations. Multiple or complementary sensory information may guide many everyday behaviours of birds, ranging from foraging for specific items to finding the way around a territory or home range, or finding the way across or between continents.

Minimal information

Sensory ecology also suggests that birds are often acting at the limits of their sensory information, responding to what might be considered minimal information, well below the detail that their senses are capable of providing. This use of minimal information should not, however, be too much of a surprise. It is something that we humans often do. However, it is something that we are rarely aware of. There are many occasions when we are prepared to act, sometimes dangerously, although we are not gaining information from within the zone in which our senses perform well and reliably. There is good evidence that birds do the same thing.

We can readily observe human performance at the limits of sensory information when we watch sports. Strength and speed may be obvious in excellent sporting performance, but usually sports people are also working at the limits of their sensory information. Catching or hitting a ball is often achieved at the very limits of what the sense of vision can provide. Gymnasts will go through their routines guided by minimal touch cues; track cyclists are often responding to minimal visual information about the manoeuvres of their competitors.

A more everyday example of humans working at the limits of information occurs when we choose to travel fast. Travelling faster than running speed is very new in our evolutionary history, and often it seems that our senses cannot match the modern challenge of speed. Driving a vehicle at speed (often 10 times faster than an average runner and 20 times faster than a walker) exemplifies an important principle of moving safely at or beyond the limits of sensory information; it can be achieved, but only in relatively simple and predictable situations.

When driving, it is sobering to reflect on the informational challenges that we face. We are very often travelling at speeds well beyond the information that we have immediately available, but we drive as though we have that information. But we can also reflect on how societies have arranged things to allow us to drive safely at speed. The answer is that roads, and the behaviours of other drivers, are engineered and constrained to be highly predictable. This is achieved through high uniformity in the standards of road engineering and in the paraphernalia with which we mark and sign roads.

Humans have developed ways to cope with these challenges of working at the limits of information through engineering solutions. However, through the construction of our artefacts we have also presented birds with many sensory challenges that go far beyond those that their senses evolved to cope with. Hence, we find that many species of birds are prone to collisions with apparently large and obvious human structures, including buildings, wind turbines, power lines, and fences. Other bird species may be faced with novel inedible 'food items' with the result that items made of plastics may be ingested, with disastrous outcomes for individuals and populations, especially of seabirds (Figure 1.3).

These are examples of birds being faced with informational challenges that go well beyond those that they evolved to cope with. But birds also perform actions that are guided by partial information at the limit of what they can detect. This is seen most obviously in species that are active at low light levels, whether it is in a forest at night, within a cave, or in deep or turbid waters (Figure 1.4). As with the case of human car driving, it seems that these behaviours are possible because they occur in highly predictable situations. This predictability can be due to the spatial simplicity of a habitat, but it can also occur in spatially complex habitats which are rendered predictable through a high degree of knowledge gained through territoriality and long residence.

FIGURE 1.3 Plastic rubbish and power lines both pose serious problems to birds of many species. Large numbers are killed annually due to collisions with power lines and other artefacts, such as wind turbines, fences and netting. Other birds die as a result of ingesting plastic items, sometimes in large numbers. There is growing evidence that these are hazards for birds because they present sensory challenges that are beyond the limits of the information that they have evolved to extract from their environments. (Photograph of plastic pollution on Accra beach, Ghana, by Muntaka Chasant [CC BY-SA 4.0]; photograph of power lines crossing the Karoo, South Africa, by Jessica Shaw.)

FIGURE 1.4 Great Grey Owl *Strix nebulosa* and Great Cormorant *Phalacrocorax carbo*: two species that exploit the resources of challenging environments in which sensory information can often be severely restricted. It seems clear that both species frequently carry out tasks beyond the limits of their immediate sensory information. (Photo of Owl by Danny Green, Cormorant by the author.)

This book

This book has an ambitious aim. It is to encourage in readers a full and detailed entry into the world of birds, to appreciate the world from their perspective, not ours. We have many different way of contemplating birds' worlds. For example, we may be observers of spectacle, admirers of evolutionary adaptations, or tickers-off of the diversity of species. All of these are exciting and enjoyable ways of appreciating birds. But there is much more to birds if we can enter their worlds by learning about their senses and thereby understanding the many ways that birds use information to guide their actions in different environments. In short, the aim of this book is to appreciate the worlds of birds through the information that their senses provide. Hopefully this book will make you feel that you can be a participant in the challenges that face different bird species as they conduct their special behaviours in particular environments.

Reflect a while on your favourite species, or on a species that particularly intrigues you. How can you piece together an understanding of their world? What do you need to know about? These birds may be living right alongside you, sharing the same environment, but in truth you and they live in different worlds, each species extracting a unique suite of information about that environment.

Through their eyes and ears, through their senses of touch, taste, and smell, and through magnetoreception, each species gains different information, so much

FIGURE 1.5 A Golden Eagle and a human in the same place and at the same time, but what each knows of that situation is quite different. Each is living in a different world bounded by their different senses. Both worlds, however, are equally real and valid. (Photo by Simon Baxter.)

so that each species is tuned into a different world. Knowing that this is the case can be both challenging and exhilarating for understanding the world in which we personally live. By reflecting upon the sensory challenges and their evolved solutions in other species, we become a little more aware of what it means to be human. We may also become a little more humble about our place in the world when we realise that what we think of as reality is just one of many realities, each reality defined by a different species, and the information provided by their suites of senses (Figure 1.5).

Chapter 2

Investigating senses

We very rarely think about what our senses are doing for us. Always working in the background, our senses are extracting a continuous flow of diverse information from the environment surrounding us. This informational flow is essential for guiding our every action. Only if we perform carefully controlled tests can we gain insights into the information that constantly flows into us. Even when we apparently 'switch off' by falling asleep, our senses are providing information vital for our wellbeing. Most of the time this information is accurate and reliable, but it can sometimes be erroneous or unreliable. Not only do our senses have limits, but uncertainties can arise from the senses themselves, and from the way the brain interprets the information that they provide. Simple tricks, such as sound and visual illusions, demonstrate that the information extracted from our senses can be fallible (Box 2.1).

That we are usually unaware of what our senses are doing might also encourage us to believe that senses are cheap to run, that the information we receive is cost-free. Certainly, compared with the energy needed to maintain our body temperature or move our bodies about, the information that we use to control our movements seems to be gained without effort. However, while our muscles and internal organs burn up calories in a tangible way, our brains are also always working hard, burning lots of calories. And most of that brain work is in the extraction and integration of information from our senses.

Every day our brains use about 20% of the energy that we consume. Nearly all of this energy is used just to keep information flowing in from the senses and for directing our actions based upon it. We like to think that 'thinking' is hard work. However, it uses very little energy compared with the effort needed to provide us with the flow of well-integrated information from our senses upon which our thoughts are based.

That the activity of our senses is metabolically expensive strongly suggests that the information they provide must indeed be very valuable. Nature's arbiter of utility and efficiency, natural selection, would have long ago weeded out most inefficiencies in information capture, and would have also weeded out any information that is superfluous. Natural selection will have honed very exactly what information our senses need to provide our brains in order to ensure our survival. The relatively high cost of running the senses, and of integrating the

Box 2.1 Illusions and reality

Visual illusions are often used to amuse. Simple drawings that can be 'seen in two ways', but never both ways at once, always bring a smile. Lines that are the same length but look to be of different lengths can bring delight and consternation. Illusions, however, are far more than tricks. Properly used, they reveal a great deal about how sensory information is interpreted by the brain. Illusions, using either vision or sounds, demonstrate that the same information provided by the senses can be interpreted in different ways by the brain. Illusions also show how the brain constantly seeks to bring order when the senses provide ambiguous or scant information.

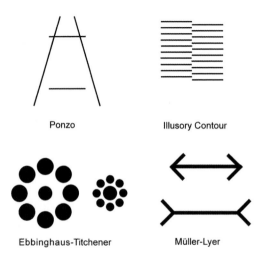

Ponzo   Illusory Contour

Ebbinghaus-Titchener   Müller-Lyer

information that they provide, will have been a constant theme throughout all animal evolution. As animals have radiated from ancestral forms to occupy different ecological conditions and exploit different resources, it should not be surprising that their senses have also diversified and become tuned to provide information for the efficient execution of different tasks in different environments.

Only very rarely are we aware of the information that we respond to at any instant. Furthermore, we are usually unaware of how information is changing continuously from moment to moment. The brain integrates and smooths things out to provide continuity of experience. Conscious reflection on what we are seeing, hearing, smelling, or feeling is slow, and the world about us changes so rapidly that we are unable to reflect on what we are responding to.

Humans are very proud of their consciousness; we like to think of it as something that makes us special within the animal kingdom. Losing consciousness is always viewed as a catastrophe, something to be regained as soon as possible. Conscious thought is, however, a cumbersome process. So cumbersome that we are unable

Illusions raise the rather uncomfortable question of 'where does reality lie?' In two of the illusions shown here (Ponzo and Müller-Lyer) the horizontal lines are of the same length. In the Ebbinghaus–Titchener illusion the two central black circles are of the same diameter. The information coming through our eyes is the same, but the brain interprets it differently. In the Illusory Contour, there is no vertical line, but we see one. So can we ever really be sure of what is out there in the world beyond ourselves? What actually happens when information received at the eyes is constantly changing? Are all interpretations correct? What happens when senses are working at their limits, when information starts to be unreliable?

The illusions presented here are particularly pertinent. They either make us see something that's not there or they give us false information. Fascinating as these illusions are, it has been shown that these same illusions have similar effects in birds. Doves too see illusory contours and make incorrect size comparisons when faced with these kinds of illusion patterns.

Showing that birds respond to illusions has involved elaborate training and testing using behavioural techniques. For convenience, this work has usually been done with doves, but other species including finches and owls have also been shown to experience visual illusions. With some illusions the birds seem to experience the illusion in a slightly different way to how we do. This illustrates that the processing of visual information in bird brains is not the same as in ours. However, the important point is that under certain circumstances birds' brains can give them rather tenuous understanding of the reality of the world about them, that their brains work hard to resolve ambiguity, and that their senses can be as unreliable as ours.

to keep up a conscious commentary on what we are responding to. Simply stating what you can see, hear, or smell at any instant skates lightly across the surface of our rich and varied sensory information. We need special procedures to probe our senses and to determine what information we potentially have available to guide our actions. Such probing is difficult to achieve in humans, but it is additionally difficult when we seek to find out what information is available to non-human animals.

The origins of investigating senses

The origins of thinking about the problem of understanding human and animal senses go back more than 2000 years in western thought. However, it is only during the last 200 years that we have gained real insight into the diversity of sensory information across species. The Darwin–Wallace ideas of natural selection, and Alexander von Humboldt's ideas about the interrelationships of organisms with their environment, provided two big ideas which now provide broad frameworks for thinking about the senses of animals. They give us many reasons for expecting that senses, and the information that they provide, will differ between species. Importantly they provide ways to account for these differences in terms of their evolution and their ecological functions. We can now appreciate why sensory information might differ between species, and how it is linked to specific behaviours and ecologies.

Epicurus and Sextus

Ideas about the challenges of understanding the sensory world of humans were first thought through and described by the Greek philosopher Epicurus (341–270 BCE). The writings of Epicurus, and the elaboration of his ideas by later philosophers, have given rise to the substantial body of thought known today as Epicureanism. Many prominent thinkers through to the present day have characterised themselves as Epicureans. Epicureanism has a number of key ideas, and prominent among them is being prepared to accept the limitations on the information that is available to guide actions.

Epicurus was among the first to write explicitly about the way that our senses place very real constraints on our overall understanding of the world. He was the first western philosopher–scientist to recognise that the information we receive changes from moment to moment. He also argued that information, even though it is concerned with the same object, is radically different depending upon the sense involved. What becomes recognised by us as a particular object is constructed from many different, and ever-changing, sources of information received from different sensory systems.

Some 500 years later, the Roman philosopher Sextus Empiricus (160–210 CE) took these human-focused ideas of Epicurus and saw their implications for all animals (Figure 2.1). Sextus argued that non-human animals are also constrained by the information that their senses provide and, crucially, that different animal species are constrained in different ways, so that they cannot possibly be living in identical worlds. This was probably Sextus' most important insight. Today we have a wealth of scientific data to support such a notion.

It might not have been unreasonable to surmise that while birds experience a different world to that of humans, all bird species experience the world in the same way. Thus, it would have been possible to explain that bird species exhibited different behaviours simply because of differences in their anatomy and bodily structures. That is, it was only because of the differences in their bills, wings, legs, etc. that birds behaved differently, not because of any differences in the information that they had about the world. Sextus argued that this was not the case, and that while species might share the same environment the information that they had about it was different.

Sextus argued that the anatomical and structural differences between species went hand-in-hand with differences in the information that each species received

SEXTUS   EMPYRICUS

Ex numismate æreo.

FIGURE 2.1 The Greek philosopher Epicurus (left) and the Roman philosopher Sextus Empiricus (right). Their ideas laid the foundations for a comparative approach to studying the senses of animals. They viewed sensory capacities as linked to the environment in which an animal lives and to the tasks that it performs. Shaped by modern ideas about evolution, their approach to understanding the worlds of animals is manifest today in the science of sensory ecology. (Image of Sextus Empiricus from Wikimedia Commons, public domain.)

to control their behaviour. Having established this as a key idea, it can be seen that in essence sensory science, and especially sensory ecology, has for two millennia been filling in the details – an enterprise that has been boosted significantly by the evolutionary and ecological frameworks that emerged in the last 200 years.

Sextus' insights and their role in the evolution of fundamental ideas about the world have been profound. This is because they lead to a quite unsettling position. They inevitably lead to the question of what the 'real' world might be like. Is there such a thing, or are there many different worlds depending upon the information available to particular species? It was questioning in this way that laid the foundations of Scepticism – the idea that we must neither accept any idea as true nor any idea as false, but we must always question. Scepticism is based upon the recognition of how difficult it is to be sure of the world when it can be based only on the information that the senses provide. Scepticism, in turn, led to Empiricism and the system of enquiry that underpins the modern scientific approach to understanding the world through experimentation and hypothesis testing.

Questions about senses: differences and dimensions

Questions about the sensory world of birds are legion. Every reader will have his or her own set of questions. Questions may be focused around particular bird species, but with almost 11,000 species, specific answers are not always available. Some questions have straightforward answers, some answers will be nuanced, and many will not have an answer ... yet. We might ask, for example, about how the ability to see details in a scene differs between species, or how visual fields might differ between species. If we know the answers to such questions, we might be able to account for a raptor's ability to capture its prey, or why species differ in their vigilance behaviour. If we know something about hearing, we can ask about the sounds used by birds to advertise their presence in a territory and whether birds can accurately pinpoint a singing rival. If we know about smell and taste, we can ask about their role in helping birds to locate profitable foraging locations.

Such questions lead to the three core challenges of sensory ecology:

1. to measure sensory performance so that it is possible to properly compare like with like across species;

2. to understand the anatomical and physiological differences that are responsible for differences in sensory capacities;

3. to propose ideas about what drives these differences from an ecological and behavioural perspective.

Devising ways of comparing sensory performance is a key challenge because we need to be confident that the same thing is being measured in different species. Despite differences in the size, general structure, behaviour, and ecology of bird species, it is necessary to get information about the same sensory capacities in each species. Furthermore, while it is possible to ask in general terms what a bird can see, hear, smell, taste, etc., each sensory capacity is highly complex (Figure 2.2).

The only way to characterise and quantify each sensory capacity is to subdivide it into particular dimensions. It is these subdivisions which can be measured and compared with confidence. For example, when investigating vision, comparisons across species are usually made with respect to a number of attributes. Prominent among these are resolution (the ability to detect detail), contrast sensitivity (the ability to detect differences in brightness), absolute sensitivity (a measure of the smallest amount of light that can be detected), relative sensitivity across the spectrum of light, colour discrimination (the ability to detect differences within the spectrum), and visual fields (a description of the space from which visual information can be extracted at any instant). Some of these dimensions of visual ability are also subdivided. For example, resolution and contrast sensitivity are found to differ when considering targets of high contrast (black and white) versus targets that contain lights of different colours (Figure 2.3).

It will also be crucial to know how performance on these dimensions of vision is influenced by ambient (overall) light levels. This will be particularly important

FIGURE 2.2 A photographic montage of bird species from a wide range of orders and families. Each species has a unique biology, and this montage could well be used to show diversity of bill structures and sizes and how they can be related to different diets and foraging techniques. However, aspects of the senses, especially vision, in all of these species have been investigated. These show that these species also differ in the information that their senses extract from the environment. Furthermore, it has been shown that these are intimately linked with both the ecology and the behaviour of these species. In short, the eyes and vision of these birds vary as much as their bills.

not only for comparing between species, but also for trying to understand how vision limits and constrains behaviour in natural environments. This is because in every natural environment, apart from underground or deep below the surface of water, light levels vary greatly through the daily cycle. When asking questions about what an animal can see, the answer will always depend upon how much light is around.

Unfortunately, this means that the answer to a question about what a particular bird might see, and how it is related to particular behaviours, has to focus upon particular dimensions or capacities within vision, and even at what time of day the bird is active. General overall statements may be difficult to make, but as in so many things there is fascination in subtle detail and nuanced answers.

Investigating all of these different aspects of vision is a tall order especially in light of how species can differ in terms of their behaviour, size, physical structure, etc. In essence it is necessary to find ways of asking animals questions and getting clear unambiguous answers about their senses. No matter what the birds may do, what they look like, or where they live, it is useful to know some basic things. What is the finest detail that they can see? What is the frequency range of sounds

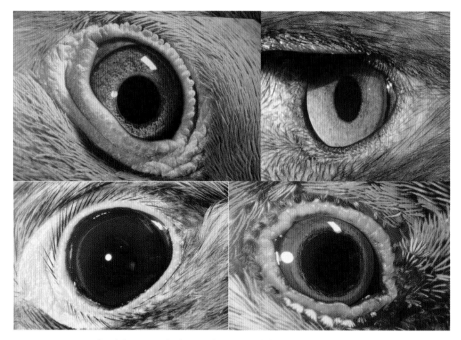

FIGURE 2.3 Four birds' eyes. Clockwise from top left: Rock Dove *Columba livia*, Golden Eagle, Orange-breasted Waxbill *Amandava subflava*, Common Kestrel *Falco tinnunculus*. Although all are built to the same basic design, these eyes differ in many aspects of detail. This includes their overall size, the optical properties of their corneas and lenses, and the structures of their retinas. As a result, the vision of each of these species is unique, and each eye provides different information about the worlds in which these birds live.

that they can hear? How finely can the animal discriminate between lights of different colour, and sounds of different frequency? Which airborne molecules can be detected by their olfactory systems? How fine is a bird's ability to discriminate the frequency and amplitude of movements of an object touching its bill tip?

Answers to such questions help put broad limits on what an animal can detect and therefore what can influence its behaviour. Unfortunately, even these questions have been answered in relatively few species, and rarely are answers available to all such questions in a single species. However, we do know enough for some general principles to have been established and some rules of thumb to have been generated.

Describing sensory performance

How can sensory performance be quantified? In essence we want to ask some specific questions. We want to ask a Barn Owl and a Starling what it can see and hear, a Sanderling what it can feel at its bill tip and taste on its tongue, a Storm Petrel what it can smell (Figure 2.4).

Sensory science is mainly concerned with these kinds of questions, with revealing sensory capacities and discovering the mechanisms that underpin them. Sensory ecology, however, takes this information further and is more interested in revealing what sensory capacities are in play in a particular situation, or in finding out how an animal uses the information that it has available to guide its key behaviours.

The aim of sensory scientists has been to manipulate just one or two parameters of a stimulus at a time, and if possible to use those manipulations to determine

FIGURE 2.4 Four bird species which differ markedly in their senses and how they use them to gain information that is used to guide their behaviour, especially their foraging. Clockwise from top left: Common Starling *Sturnus vulgaris*, Barn owl *Tyto alba*, Leach's Storm Petrel *Oceanodroma leucorhoa*, Sanderling *Calidris alba*. These birds differ in their vision, hearing, sense of smell, and sense of touch. Furthermore, each species relies upon a different primary sense to guide its behaviour. They also differ in how they combine and complement information gained through different senses. In short, each bird lives in a different sensory world. (Photo of Starling by Pam P. Parsons [West Bay Dorset, via Flickr as Pam P Photos], Barn Owl by Graham White [CC BY-NC-SA 2.0], Storm Petrel by C. Schlawe [public domain], Sanderling by J. J. Harrison [https://www.jjharrison.com.au, CC BY-SA 3.0].)

the limits of sensory performance. If this can be done in a comparable way across a range of species then we should be able to say with some confidence that this species is more sensitive than that one, or this species is able to gain information over a wider range of parameters. These are often the kinds of things that birdwatchers, journalists, and TV documentary makers are keen to know about.

I am often asked whether this or that species is 'better' or 'worse' than humans. But the biologically relevant question is whether the species is better or worse than the species that it is competing with. What is important to a sensory ecologist is using this information to understand what information an animal has available to guide its behaviour in real-world tasks, and understanding how information from different senses might be integrated.

It does not matter whether an animal's senses are 'better' or 'worse' than humans. Humans are, after all, just one species, adapted to living in certain types of environments through the conduct of particular behaviours, so comparison with humans may not be important. The desire to compare another species' sensory performance with humans is born of an anthropocentric view of the world. It is a viewpoint which provides a strong pull. I try to resist it, but we shall not be able to escape from it completely in this book.

Measuring senses in a similar way in different species does have great value for comparative studies and also for helping to understand the basic mechanisms that underlie a sense. For example, if differences are found between species in their ability to see detail, and systematic differences are also found in the structure of the eyes' optics or retinas, then it is possible to start piecing together an understanding of basic mechanisms. This in turn allows the possibility of being able to predict what another species might be able to see just from knowledge of its eye structure.

Sensory thresholds

Limits of sensory performance are defined as *absolute thresholds*. These are measures of the minimum amount of a particular stimulus that can be detected. In the case of sound, this will be the smallest amount of air disturbance that the hearing system can detect. In vision, it is the lowest number of photons received per unit time that can be detected. In the chemical senses (taste and olfaction), it is the lowest concentration of molecules of a particular substance that can be detected. In mechanical senses (for example, touch sensitivity of a bird's bill) it is the smallest displacement of the detecting surface.

Absolute thresholds are particularly interesting because they allow simple and direct questions to be posed. Which species has the most sensitive hearing? Which species has the most sensitive eyes? However, comparing thresholds is not as straightforward as one might hope, and answers are rarely clear-cut. This is because thresholds differ not just between species but between individuals.

Like all measures of performance, absolute sensory thresholds tend to be normally distributed when the results for individuals are collected together. This means that the absolute threshold for a species is best described statistically as a mean value with variation around it based on a sample of individuals. Between-individual differences are often particularly notable if age comes into play. A change in sensitivity is usually part of the ageing process in birds, as much as in humans. Therefore, to compare species it is necessary to refer to mean differences, and to be aware that individuals will have sensitivities below or above this mean.

A simple analogy is to consider what the answer might be if you asked how fast humans can run. Every able-bodied person can run, but there is no surprise that the running speeds of people will differ, often markedly so. The answer is that there will be a mean speed for human running performance but around the mean there will be a wide range of speeds. It would then make sense to compare the average running speeds of humans with another animal species, and while there might be a difference in the averages there may well be some overlap in the performances of individuals of the two species.

It is the same with absolute sensory thresholds: differences between individuals do occur. Also, an individual's thresholds can change owing to a range of factors, including their motivation to participate in the investigation. So, although absolute thresholds are both important and interesting, and will feature a number of times in this book, it will always be necessary to express caution, or at least bear in mind that sensory thresholds are mean values based on a sample of individuals representing a particular species.

Relative sensitivity within a sense

Clearly, the sound and light that an animal is able to detect can vary in their total energy. We can see very bright as well as dim lights, hear very loud sounds as well as very quiet sounds. The ability to detect these stimuli indicates that there is a wide dynamic range to sensory performance. Absolute thresholds measure only the ability to detect the lowest amount of energy for a particular type of stimulus or sensory dimension.

As well as the wide dynamic range, most types of natural stimuli can also vary along a number of dimensions. For example, the vibrations of air molecules that we detect as sounds can have different frequencies, and it is these different frequencies that humans describe as sounds of different pitch. Light also varies in frequency, though we more often describe it in terms of its wavelength. Our visual system detects the different wavelengths as lights of different colours.

In each animal species there are likely to be different absolute thresholds for each frequency of sound or each wavelength of light. This means that a complete description of a bird's vision or hearing requires knowledge of thresholds across a wide range of stimulus frequencies. For light these are presented as *spectral sensitivity functions*, while for sounds they are presented as *audiograms* (Box 2.2).

Perhaps the most familiar of these are the differences that can occur in humans in their sensitivity to sounds, in our individual audiograms. If we have reason to

Box 2.2 Audiograms and spectral sensitivity curves

Light and sound have an important feature in common. Both can be described as waves. These disturbances are comparable to the peaks and troughs of a wave that we might watch travelling across the surface of a pond.

The distance between two peaks of a sound wave is referred to as the wavelength of that sound. Since the speed of travel of sound or light through a particular medium is constant (although sounds travel four times faster through water than through air, and even faster through solids such as metals and rocks) the wavelength bears a simple relationship to the frequency with which the waves pass by. It is this frequency that is usually used to describe one of the key properties of sounds. The unit of frequency used is hertz (Hz), the number of wave cycles per second, but sounds are commonly described in kilohertz, 1000 wave cycles per second.

The frequencies of different sounds vary, and it is these frequency variations that ears detect. We refer to sounds of high and low frequency as being of high and low pitch.

The relative heights of the peaks and troughs in a wave can also vary so that two sounds of the same frequency can contain different amounts of energy. These differences are detected as louder or quieter sounds.

The full spectrum of sounds in nature is remarkably broad, but any animal's ears can detect only a portion of all possible sound frequencies. Across a range of frequencies an ear detects all sounds but sounds outside that frequency range are simply not detected. Knowledge of the upper and lower limits of the hearing range of a species is an important part of describing their biology. Different animals can hear sounds across markedly different frequency ranges, but no species can detect all possible sounds. Some species detect sounds in the highest frequency ranges, other in the lower ranges, and for others hearing is in a mid-range of frequencies.

Within the frequency range of an animal's hearing, sensitivity to sounds is not the same across all frequencies. Some sound frequencies can be detected at lower intensities (the sound waves have low amplitude), while others can be heard only at higher intensities (waves of high amplitude). When hearing sensitivity has been determined across the range of frequencies that an animal can hear, an *audiogram* can be constructed for that particular species.

At a glance audiograms convey a lot of information about the average basic hearing abilities of a species. They define the upper and lower limits and also show which frequencies can be detected at high amplitude and which can be detected at low amplitude. Audiograms generally show a broadly 'U-shaped' function. This indicates that there is lower sensitivity (high-amplitude sounds are needed for their detection) to both higher- and lower-frequency sounds, and that the highest sensitivity (sounds of low amplitude can be detected)

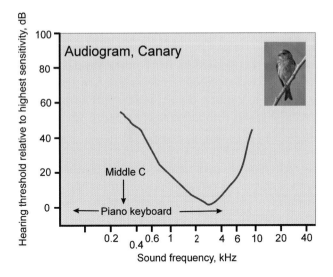

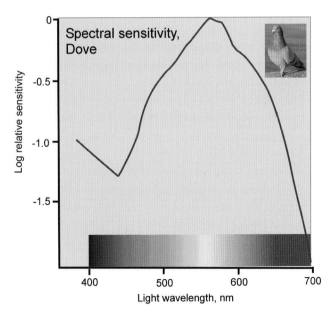

occurs in a middle range of frequencies. However, audiograms are typically not symmetrical, with rapid changes in sensitivity as a function of frequency in some parts of the hearing range, and less rapid changes in others.

The diagram here shows the average audiogram for Atlantic Canaries *Serinus canaria* (the common cage bird). It shows that maximum sensitivity to sound occurs at relatively high frequencies, the sounds of notes at the higher end of the piano keyboard, but above those frequencies sensitivity drops dramatically.

At the low-frequency end, it looks as though a Canary will not hear the lowest notes of a piano, or at least they would have to be relatively loud to be detected.

If audiograms are available for a number of species, then differences in their hearing can be comprehended readily by comparing audiograms. This makes it possible to get a clear and rapid understanding of how the hearing of species differ one from another. Because of this, audiograms have become a valuable tool for characterising and comparing hearing across species throughout the animal kingdom.

Audiograms are also used as a clinical tool in humans. They are used to characterise different types of hearing loss by comparing an individual's audiogram with a normal or average audiogram for humans. An individual's audiogram is likely to change with age. Hearing loss at higher frequencies usually occurs with increasing age, and this readily shows up in an audiogram. This can be used to define the best characteristics of a hearing aid that can be recommended for a particular person.

Diagrams showing the differential sensitivity to light of different wavelengths are known as *spectral sensitivity curves*. Like sound, light can also be described as a wave phenomenon. However, rather than referring to the frequency of the wave it is usually more convenient to describe light by reference to its wavelength. The distance between the peaks and troughs in light waves is very short compared with sound waves.

The actual wavelength of light is too small to comprehend readily. However, by using the nanometre as the unit of measurement the numbers become manageable. A nanometre (nm) is one metre divided by 1,000,000,000, and

have our hearing tested the results will be compared with a normal or average audiogram for humans. This shows how our own hearing may be more or less sensitive to particular frequencies than the hypothetical 'normal' individual. However, an inevitable process of ageing is that people start to lose their sensitivity, particularly to higher-frequency sounds. This is referred to as differential hearing loss, and the details of this loss describe how much our own hearing may have changed over time. However, it is worth bearing in mind that these changes can be the result of natural processes, although they can result from disease or physical damage caused by exposure to loud sounds.

When audiograms are compared across a wide range of animal species (mammals, birds, reptiles, and fish) very notable differences in sensitivity to sounds of different frequencies are found between these main animal taxa, as well as between individuals within a taxon. Some animals are able to detect sounds of very low frequencies, others are able to detect sounds with very high frequencies. The same applies to vision, where some species are able to detect light of short wavelengths and others light of relatively long wavelengths. Sensitivity to these using this unit it is possible to refer to visible light (for humans) as falling within the range of approximately 400–700 nm (Figure 2.5).

Within the visible spectrum of an animal, sensitivity to light is not the same across all wavelengths. Light at some wavelengths can be detected at lower intensities, while others can only be detected at relatively higher intensities. When visual sensitivity has been determined in a particular species at a range of wavelengths a *spectral sensitivity curve* can be constructed. The curve for the Rock Dove (Feral Pigeon) *Columba livia* at daytime light levels is shown here. It indicates that the eye is most sensitive to light in the yellow-orange wavelength range and that sensitivity falls rapidly in the reds and in the greens and blues, but there is a slight rise in sensitivity in the violet range.

At a glance spectral sensitivity functions convey a lot of information about the average basic vision of a species. These functions generally show a broad domed shape. This indicates that there is lower sensitivity (high-intensity lights are needed for their detection) to both longer and shorter wavelength lights, and that the highest sensitivity (lights of lower intensity can be detected) occurs in a mid-range. However, the position of the peaks in sensitivity and the shapes of spectral sensitivity functions are usually not symmetrical, and in some animal species more than one peak can occur.

If spectral sensitivity functions are available for a number of species, then differences in their vision can be comprehended readily by comparing the functions. Therefore, they are a valuable tool for characterising and comparing vision across species. They are also used as a clinical tool in humans to detect different types of vision loss.

shorter wavelengths of light is referred to as vision in the ultraviolet part of the spectrum. In hearing, sounds of low and high frequency are referred to as infra-sounds and ultrasounds.

These infra- and ultra- labels are derived from comparisons with the range of light and sound that humans can normally detect. Although in an anthropocentric world it seems natural to use humans as the base comparator species, seeing and hearing outside the range of humans is nothing special, it is just different. Indeed, we might well ask why human hearing is stuck in the middle and does not reach these ultra- and infra- ranges. Many of the mammals that we share our everyday lives with hear sounds that are well outside the range that we can detect (Figure 2.5).

Costs and trade-offs in senses

Humans are all too keen to think that they 'know' the world, that the world is as they perceive it through their senses. However, even the above brief mention of

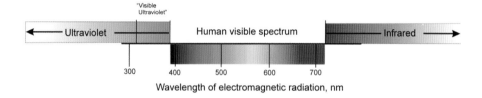

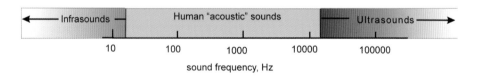

FIGURE 2.5 The spectrums of light and sound, showing the general range of frequencies and wavelengths that humans and other animals are able to detect. Infrasounds have frequencies below those that humans can hear, ultrasounds have frequencies higher than those that can be detected by humans. However, there are many animals species which can detect sounds within the infrasound and ultrasound frequency ranges. Similarly the electromagnetic, light, spectrum has a range of wavelengths that humans detect as spectral colours (the colours of the rainbow) from violets to reds, but some animals can detect light at shorter wavelengths, in the ultraviolet part of the spectrum, and some can detect light in the infrared part of the spectrum.

infrasounds and ultrasounds, of infrared and ultraviolet light, tells us that the world contains far more information than we can directly receive. By definition these are sources of information about the world that we cannot directly access. These are sounds that we cannot hear, lights that we cannot see, but other species can. This means that it is arrogant to believe that humans should be the comparator of all things – but it is equally true that no organism can fulfil that role. Put simply, it is not possible for any one species to be able to detect everything that is going on in its environment. It is not possible for any one species to 'know' comprehensively how the world actually is.

Trade-offs between senses

The world of any one species is no more important or special than that of another. All sensory worlds have equal importance. They have been shaped by natural selection to extract information for the efficient conduct of the life of each species. For individual species there will be important constraints on how their sensory organs can perform. This is because there are costs and trade-offs in sensory capacities within a sense, and also between senses. The trading off of information between senses is something that will be discussed a number of times in this book. One particularly dramatic example is found in some ducks which, unlike most other birds, have comprehensive vision of the world about their head. This is only possible, however, because their foraging has become controlled by touch and taste information, so that they do not need to see where their bill is. This has resulted in a trade-off between vision, touch, and taste that has given rise to particular diets and behaviours.

An important constraint on how animals detect their environments comes from the metabolic costs of operating different sensory systems. Vision is particularly costly. Not only are eyes demanding of support and protection in the skull, but their actual running costs are high. There is a rapid and constant turnover of materials and large amounts of neural processing are necessary to extract information from visual input, and neural processing is demanding of energy. Eye size is a fundamental factor in both visual resolution (the amount of detail that can be extracted from a scene) and sensitivity (the minimum amount of light necessary for the extraction of information). As a general rule the larger the eye, the higher its sensitivity and resolution.

The eyes of most birds are small, but there are plenty of species that have large eyes – for example owls, albatrosses, raptors, hornbills, and penguins. In all of these species larger size would seem to be the result of natural selection for either high sensitivity or high resolution, or even both. But not all nocturnal species have large eyes. We might reasonably predict that large eyes could have easily evolved in kiwi species, because they are flightless and weight should not be a problem – but in fact the eyes of a kiwi are similar in size to those of a small passerine. The answer to this apparent paradox lies in the fact that kiwi conduct many tasks guided by information derived from non-visual senses, most notably smell, hearing and tactile cues from the bill tips. This is another striking example of how information from one sense can be traded off or complemented by information from another. In the case of kiwi, the result of these trade-offs is that their sensory world, their reality, is far removed from those of other nocturnal birds.

Trade-offs within a sense

When attempting to understand the behaviour of particular species the above examples of complementarity or trade-offs *between* the information received from different sensory systems are particularly important and fascinating. However, significant trade-offs also occur *within* a sensory system. In fact, compromises and trade-offs within a sense should be considered the norm. This reflects the simple truth that within a particular sensory system it is not possible to collect all of the information that is potentially available.

Trade-off within a sense is clearly seen in the relationship between visual *resolution* and *sensitivity*. At the very limit, both resolution and sensitivity are determined by the quantal nature of light and by noise within the nervous system. This trade-off is evident most dramatically in the fact that resolution always decreases as higher sensitivity is gained. We experience this every day of our lives: as light levels naturally fall, and our eyes become more sensitive to dim light, we accept that there is less detail in a scene. But the detail has not gone away, it is still there, it is just not detected by our eyes, which are adapted to detect the reducing number of light quanta in the environment.

FIGURE 2.6 The trade-off between high sensitivity and high resolution is found in all vision and imaging devices, including cameras and eyes. It arises from fundamental constraints imposed by the quantal nature of light. The trade-off is exemplified here by two bird species. In Short-toed Snake Eagles *Circaetus gallicus* vision has evolved to provide high resolution but low sensitivity, while in Tawny Owls *Strix aluco* high sensitivity is achieved but resolution is low. Hence, while the vision of a Tawny Owl is suitable for activity at low light levels it has low resolution and cannot detect fine details. On the other hand, eagles achieve high resolution but have low sensitivity and so they tend to cease activity as light levels fall towards dusk. (Photo of Short-toed Snake Eagle by A. Román Muñoz Gallego, University of Malaga.)

Our loss of spatial details with decreasing light levels is not just a quirk of our vision. It is because it is a fundamental constraint on any vision system, including cameras. It is the very physical nature of light, its quantal nature, that precludes high visual resolution at low light levels. An eye that has evolved to achieve high sensitivity is unable to detect fine spatial information at low light levels, but neither can a highly sensitive eye readily achieve high resolution when there is a lot of ambient light. Life is full of compromises, and that is certainly true both within and between different sensory systems. Natural selection has worked on these trade-offs and fundamental constraints to shape sensory information optimally for the conduct of different tasks, in different environments, by different species (Figure 2.6).

A unique property of vision

The trade-off between resolution and sensitivity exemplifies a unique and important aspect of vision. Within the range of light levels in which vision normally operates the information that it can extract varies markedly with the amount of the stimulus (light) that is available.

All other senses effectively provide the same range of information within their normal operating ranges. Across a wide range of sound levels, or concentrations of chemical compounds, or physical components of a touch stimulus, the sensory systems are able to extract more or less the same information. In vision this is not the case. This is because the information that vision provides is based primarily upon spatial resolution (the ability to see details in a scene or to say exactly where something is placed in the field of view), and resolution changes profoundly as light levels change (Box 2.3).

The change in resolution with light level is significant. Unfortunately, this makes it very difficult for an observer to determine exactly what visual information could

Box 2.3 Resolution and light levels

The effect of light levels on spatial resolution is significant. Acuity is often measured or estimated at high daytime light levels, but in a few bird species acuity has been measured across almost the full range of naturally occurring light levels (See Box 2.4 for a discussion of how natural light levels vary depending upon the elevation of the sun and moon). In some species acuity has been determined across the narrower range of the light levels that occur in daytime, and in some species across the daylight–twilight range.

This graph brings together data for a number of species and shows some key points about the effect of light levels on acuity. In all species shown here acuity has been determined using behavioural training techniques and gratings, the kind of technique depicted in Figure 2.7.

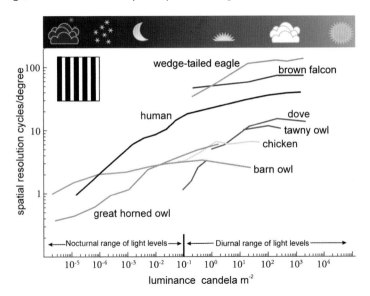

Although the maximum acuities of these different species are significantly different (for example, the maximum acuity of the eagle is 60 times higher than that of the dove) it is also clear that in all species acuity decreases considerably as natural light levels fall. The only species in which there is not a steep decline in acuity is the Western Barn Owl *Tyto alba*, but even in Barn Owls acuity shows a significant fall as light levels decrease to the lower ranges of night-time. In some species the decline is particularly steep, and this may be a significant reason why some daytime birds, such a doves, go to roost as light levels fall. If we wish to pose questions about how vision is used to control natural behaviours in any species, it is necessary to bear in mind that natural light levels have important effects on visual abilities, especially acuity.

be available to a bird at any one moment. For example, we may have knowledge of what a bird can detect at one particular light level, but it will not be the same at another light level. Furthermore, colour vision is an important mechanism that enhances spatial resolution for a wide range of natural targets. However, colour vision functions only at high (daytime) light levels. As light levels drop to those of natural twilight and below, colour vision no longer functions and so an important source of information is lost. Furthermore, even if we discount the contribution of colour vision and just consider the ability to discriminate detail in black and white, spatial resolution decreases markedly with light levels.

At any one location naturally occurring light levels may change over a range of at least a million-fold (10^6) between maximum sunlight and moonlight. It is remarkable that our eyes can function across this whole natural range of light levels. Even more remarkable is that on moonless nights the range is extended

Box 2.4 Natural illumination sources and light levels

During the day (when the sun is above the horizon), under clear skies light levels vary by about 100-fold from sunrise to full overhead noonday sun. The coming and going of cloud cover can extend this range to 1000-fold. During the night, however, light levels can be much more variable. This is because the main source of light, sunlight reflected by the moon onto the earth, varies not just with the elevation of the moon, but also with the moon's phase, which changes on a monthly cycle. This results in night-time light levels that can vary by 1 million-fold between sunset and starlight on a moonless night. On top of this there is a further potential 10-fold in variability brought by cloud cover.

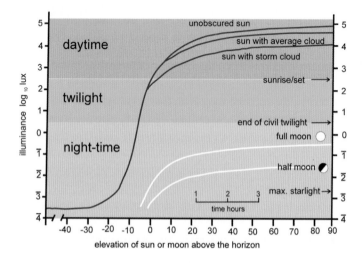

downwards by a further 100-fold, and if we take into account how the presence of clouds and tree canopies can further reduce ambient light, then the total range of light levels in which an eye can function varies by a factor of 10^{11}. That is a huge dynamic range for any detector (Box 2.4).

While an eye can operate throughout this light range, the information that it can provide changes very markedly. As explained in the next chapter these changes are not trivial, the ability to see details in a scene can be dramatically different depending on whether we are considering day or night, bright sunlight or moonlight. This means that whenever we consider what a bird might be able to see in a particular location and how it can use that information to guide its behaviour, it is always necessary to consider the ambient light levels. We need to ask about the time of day, and to consider whether colour vision might be available, or not.

This figure captures much of the huge range over which light levels naturally vary and why. It shows how natural levels of illumination at the earth's surface depend upon the elevation of the sun and moon, and upon the phase of the moon. The amount of light from these sources changes continually over the daily cycle, and over a very large range. (Note that the scale of illumination is logarithmic, which means that light levels change over 10-fold between each digit of the scale.)

Although this basic pattern occurs around the globe, it differs in detail with latitude and time of year. The basic unobscured sun curve shown here is for latitude 50° (approximately southern England) at the time of the summer solstice. Nearer the equator the rate of change of illumination levels each day is more rapid and at higher latitudes it is slower, which means that twilight (the period of transition between daytime and night-time) is shorter or longer than shown in the figure. Close to the poles light levels hardly change on a daily cycle but they do change systematically from day to day. At the equator the daily pattern and amplitude of light-level changes hardly varies across the year. Clearly, the pattern of natural light levels is significantly different across the globe and across the year, and it is within these patterns that the vision of animals has evolved.

A vegetation canopy, of course, reduces illumination at ground level. The exact amount of reduction depends upon the density of trees and their species, but a good rule of thumb is that a woodland canopy in leaf will reduce light levels by about 100-fold. This means that all of the light levels shown in the diagram are about 100 times (2 log units) lower inside a wood compared with outside it. As discussed later (Chapter 9), this can be significant when considering the challenges faced by nocturnally active birds living in open habitats as opposed to closed canopy woodland habitats.

Measuring senses

Early attempts to quantify the sensory capacities of birds relied upon anecdotal observations of behaviour, or at least behaviour that was not well controlled. These attempts typically also involved imprecise control of a testing stimulus, or perhaps no control at all, but relied on natural situations. This approach led to some rather over-the-top estimates of sensory abilities. These include some fantastic claims for the visual sensitivity of owls and the visual acuity of raptors, and even the wholesale denial that most birds have a sense of smell.

The use of well-controlled stimuli and systematic techniques for controlling behaviour now give quite different insights into the information available to birds. For example, we now see more modest estimates of visual sensitivity and a growing awareness of the importance of olfaction in a wide range of bird behaviours. Unfortunately, the results of the older anecdotal observations still linger in the literature and on the internet, and many people seem reluctant to give up old assumptions about 'super-senses' that these early studies seemed to support.

It is not difficult to understand why people favoured old interpretations based on anecdotal observations. They often squared with everyday observations of our own sensory capacities, and they often built upon myths and legends about the place of animals in the world and ideas about what it meant to be human. With a general decline in the potency of those myths and legends (including mainstream religious ideas) we are generally not so sure about what it means to be human or of our place in the world. More careful assessment of what humans can see, hear, smell, etc., compared with knowledge of these sensory abilities in other animals, now plays its part in helping to understand our position in nature.

Finding out about our own senses has proved to be a difficult task, requiring careful procedures and controls. It is more difficult to apply those techniques to ask the same questions in similar ways of non-human animals. It seems relatively easy to ask another person what they can see, hear, smell, or feel, but even then we want to know just how carefully controlled were the stimulus conditions. We also need to know whether the person was fully attentive to the task, especially if the task became increasingly difficult as the stimulus got closer to the limit of a sense's abilities. Did the person really try to see the finest detail? Hear the faintest sound? Or did they give up on the task before reaching the limits of their performance?

Animal psychophysics

The branch of science that investigates questions about the relationship between stimulus conditions and how they are perceived became known as 'psychophysics'. The term was coined 160 years ago by Gustav Fechner, who is credited as being the first experimental psychologist. The techniques developed in his early studies in psychophysics put in place a set of procedures for characterising sensory performance that persist to the present day. However, it was not until the development of ideas about 'conditional' responses by Ivan Pavlov in Russia, and later by B. F. Skinner in America, that researchers found ways of establishing 'animal psychophysics'. Using these techniques, it became possible to combine carefully controlled stimulus conditions with consistent and reliable behavioural responses to ask an animal about what it can see, hear, taste, smell, etc.

The pioneering work in animal psychophysics used the techniques of operant conditioning developed by Skinner to train animals to make specific responses when a particular stimulus or set of stimuli is presented. Remarkably, it was found that animals could be trained to respond to a very wide range of arbitrary stimuli that really had no place in the natural world in which the animals had evolved. Once the animal has learned to respond in this way to a novel stimulus it is then possible to systematically make changes to a stimulus and establish the limits of performance. Hence it becomes possible to answer such questions as how sensitive an owl's vision is, or how sensitive a pigeon is to the smell of particular compounds, and to be confident that answers are reliable and comparable between individuals and across species.

Properly controlled animal psychophysics studies involving trained animals are the benchmark for understanding animals' senses. However, not all animals can be trained easily using operant conditioning and a great deal has been learned from less systematic, often field-based, studies. Such studies often give only an indication of what a bird might be able to detect, and they often lack accurate quantification. For example, a lot has been learned in recent years about the use of the senses of smell and taste by birds. Knowledge is now being gained about which smells are more salient or important to different species, but it has not yet been possible to determine just how sensitive birds are to particular compounds.

Two-choice discrimination tasks

The most rigorous technique used in psychophysical investigations of birds usually involves a small number of individuals of a particular species being trained initially to make a choice between two quite different stimuli which are presented simultaneously. That is, it involves a 'two-choice discrimination' and trains the bird to indicate that it can differentiate between two stimuli. It is then a matter of altering those stimuli so that they become more similar until a point is reached where they cannot be told apart. This then defines the limit of a particular sensory ability.

The technique is perhaps most easily described by the example of an investigation of the limit of visual resolution. In this the investigator is seeking to determine the smallest spatial detail that a bird can see. Initially the bird is presented with two panels placed side by side, one showing a pattern of black and white stripes (a grating) oriented vertically, the other showing the same pattern oriented horizontally. The panels will be of the same size and brightness; all that differs is the orientation of the stripes. The birds will have already become used to taking small food items from a hopper or some other device in the area where the stimulus panels are displayed. Today the patterns might be shown on a computer screen or back-projected onto a panel, but previously stripe patterns on photographic negatives or printed on cards were used (Figure 2.7).

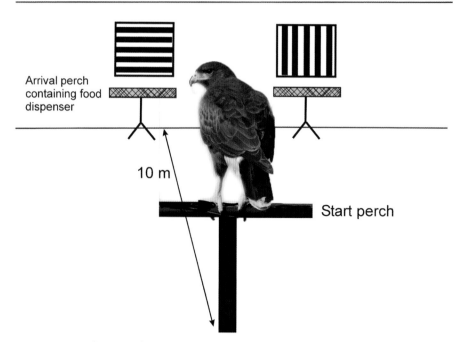

FIGURE 2.7 A schematic drawing of a setup used to determine visual acuity in a Harris's Hawk *Parabuteo unicinctus*. The bird is given a two-choice task. It has been trained to wait on the perch until a pair of patterns is shown on the panels, which are 10 m away. As soon as the bird leaves the perch the patterns are switched off and the bird flies to the perch of its choice. If it flies to the perch in front of the 'correct' panel it receives a small food reward. The bird is observed remotely so that humans cannot influence its behaviour. Many such trials are conducted and the width of the stripes on the panels is altered in a random sequence. This allows the performance of the bird with respect to the width of the stripes on the test gratings to be measured, and a threshold stripe width can be determined (see Figure 2.8).

By the time the investigation gets to this point, the person running the investigation will have got to know the birds well, but will not be in direct contact with the birds. The birds' behaviour will be remotely observed, so that the investigator cannot inadvertently influence the birds' choices. The way the investigator influences the bird is through the presentation of food items which are paired with the presentation of only one type of stimulus panel, for example, the one with horizontal stripes. If the bird is hungry it will usually learn rapidly to move towards, or peck at, the panel with the horizontal stripes, as long as it reliably receives a food item each time it does so.

I have done vision investigations with doves who will peck directly at the panels for grain, owls that will peck at a bar beneath a panel to be rewarded by a small piece of meat, and with Great Cormorants who will swim up to a panel underwater in return for a small fish dropped to them through the surface. Some species can be trained to make their choice of panel from a fixed distance and then fly, swim, or walk towards the panel along a runway or swim-way. Owls seem particularly amenable to running, while diurnal birds of prey will readily fly from a start perch and land on a perch placed just in front of a panel.

In instances where flying, running, or swimming is involved, the bird must not be able to change its choice once it has passed a certain point. This ensures that the bird's decision is always made a fixed distance from the panel. The use of a fixed distance is important for calculating the visual size of the object as seen by the bird and hence for defining its best performance. Depending on the species, it might be better for the bird to make a choice between relatively widely striped gratings at a long distance and fly to the panel (for example, a hawk), rather than choose between finer stripes at a closer distance (for example, a Budgerigar *Melopsittacus undulatus*). It is a matter of working with the bird's natural behaviours to enable it to perform the choice task readily and hence reveal the ultimate limits of its vision.

Once this link between a target panel and the reinforcing food has become established, it is then a matter of starting to manipulate by small increments the task that the bird faces. The left–right positions of the panels will be changed randomly between trials, the light levels of the panels will be changed randomly over a wide range of light levels, and the widths of the stripes in the grating pairs will also be varied between trials. The point of these variations is to ensure that the bird is responding exclusively to the orientation of the stripes. In other words, that the bird has really learned that it is the orientation of the stripes that it should be attending to.

Training and testing

This type of simple discrimination is likely to be learned robustly by most birds. Once trained, birds will make the correct choice at least 90% of the times that a pair of stimulus panels is presented. In many cases mistakes are never made, the birds being consistently 100% correct. By ensuring that the birds gain most of their daily food intake in these sessions, and that the sessions take place at the same time every day, the birds will be highly motivated to respond when they come to a training session. Also, it is best to give a bird a training or testing session every day, no time off at weekends.

Depending on the species and on the individual birds, this initial training may be completed in a couple of weeks, but it can take a couple of months. The investigator has to be alert to the possibility that the birds will learn to solve the problem posed by using a cue other than the one that the investigator intends. When birds are being trained to make these kinds of visual discriminations they can be particularly alert to sound cues which might also reliably indicate where to go to get food. So, for example, a click or rumble associated with changing the positions of the panels between left and right could well be learned, and the bird may then ignore the orientation of the stripes. This is perhaps not too surprising. After all, from the bird's point of view the task is to learn how to get its food reliably, and any cue that is consistent is as good as another, and sounds are very important to birds. So if the investigator introduces more than one reliable cue the bird could latch on to any of those that are available.

Once this initial training has settled down and the investigator can be sure that the bird is responding only to stripes of the correct orientation, the actual investigation of sensory abilities can begin. The first stage is to start presenting the birds with pairs of stripes of the same width but to change stripe width between trials. On some trials stripes will be very wide and the difference between the panels' orientations will be obvious to the bird. However, if the stripes of the grating pattern are made very narrow the task will be difficult, as the fine stripe pattern cannot be determined and the panel looks a uniform grey.

The result is that on some trials the birds may make a mistake, and errors will start to systematically creep in. If the stripes are so narrow that they are below the birds' threshold and they truly cannot ever tell the panels apart, then random behaviour would be expected, and the birds would get the task correct on only 50% of the trials. At intermediate stripe widths, however, the birds may get the discrimination correct on 70% or 80% of the trials. Over time, by mixing up trials with different stripe widths, a bird's motivation to keep responding even on difficult trials can be maintained.

The uncertain threshold

When enough trials are accumulated a relatively stable relationship between error rate and stripe width will emerge. This relationship is called a *psychophysical function*, and the point at which a bird gets 75% of the trials correct is considered the *threshold* of sensory performance (Figure 2.8). So, after all the training and testing, it is this 75% correct stripe width that the investigator is trying to determine, not what the bird can see 100% of the time.

The important point to note is that the threshold is defined statistically. It is the stripe width at which on average the birds get the choice right 75% of the time (Figure 2.8). This statistical definition reflects the nature of sensory thresholds. The limit of performance can vary from moment to moment and from day to day, depending on the physiological state of the observer and on their motivation. Keeping up a bird's motivation is very important because these investigations can last many months and the bird needs to be fit, healthy, and equally motivated throughout that period.

Once a threshold for visual acuity has been determined for a group of birds, an average value can be calculated as representative of that species. Individuals will differ, and so in any sample of birds there will be individuals above and below the average determined for the species. But it is usually the average value that is used in any comparisons between species; the best performance is no more representative of the species than the worst.

Elaborating the task

Determining an average acuity threshold is likely to be just the beginning of a much longer series of investigations. With well-trained and motivated birds, and highly motivated investigators, it is possible to repeat the investigations at

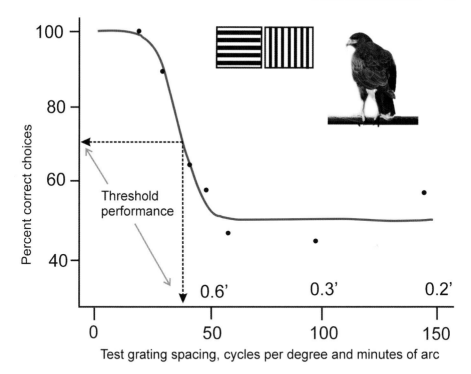

FIGURE 2.8 An example of a psychophysical function of a Harris's Hawk. Performance over a large number of trials has been accumulated on a two-choice task using the kind of setup shown in Figure 2.7. The average percentage 'correct' performance for panels showing stripes of different widths is accumulated. The black dots on the graph show the bird's actual performance for different stripe widths and the red curve is the line of best fit to those data points. With wide stripes the bird is correct on nearly every trial, while with very narrow stripes its choices are random (50% correct). It is in the region between these two extremes that the threshold of the bird's acuity lies. The region where the birds is 75% correct is usually taken as the threshold performance. In this example 75% correct would occur with stripes that are about 0.7 minutes of arc wide. Similar data are gathered for a number of birds using the same technique, and these are averaged to give an acuity estimate for the species. In the case of Harris's Hawk, the published average acuity is 1 minute of arc (29 cycles per degree).

different light levels or repeat them changing the contrast between the stripes. It is usual to start the investigation at high light levels, equivalent to daytime, and use grating patterns that contrast very highly. But many of the more interesting questions about what information is available to birds in their natural environments occur when light levels are not those of midday and when patterns are not highly contrasting. Most real-world tasks involve detecting objects that differ by shades of grey.

By systematically altering both light levels and contrast it is possible to investigate how a bird's acuity will differ between daylight, twilight and night-time, and how its ability to see details changes as contrast reduces. Actually, getting birds to continue working at very low light levels and at low contrasts takes a lot of ingenuity. This is because when tasks start to get difficult the bird may simply stop responding and hunker down until the task gets easier. Therefore, getting runs of thresholds for a wide range of light levels and contrasts can be very difficult. As a consequence, most studies tend to report acuity at a single light level and contrast, or within a narrow range. One further problem of working at low light levels is that the birds must be allowed to 'dark adapt' to the required low light after being placed into the apparatus, and the light levels need to stay the same throughout a session. This can add greatly to the time a session takes and decrease the motivation of the bird to respond.

A disconcerting experience

If conducted properly, the procedures described above will produce robust results that allow comparisons between species whose sensory thresholds have been measured in similar ways. Should the investigation call for it, it is quite possible to put a human in the same situation as the birds and for them to do exactly the same task. This has been done a number of times, and I can personally attest that measuring one's own threshold in this way is an exacting task, especially when the patterns presented are close to threshold.

There is a range of stripe widths that can be considered 'disconcerting'. When looking hard and comparing the patterns in this range, it seems impossible to make a decision – but if you look and respond quickly then you may be correct more times that you are wrong! It seems like guessing, but clearly there is a little information that makes it more than a guess. This is when visual discrimination is around the threshold level, in the 75% correct zone. It is no wonder that individuals, humans as well as birds, can lose motivation and even stop responding when faced with a choice close to threshold.

Measuring other sensory dimensions

It is quite easy to imagine how the whole procedure described above can be adapted to investigate many other sensory questions. Instead of striped patterns, the stimulus panels could be uniformly lit with white light paired with an unlit panel, or with lights of different colours. Using these and adjusting light levels, it is possible to measure the minimum amount of light that can be reliably detected; that is, the absolute visual threshold can be determined. That too is a disconcerting task close to threshold.

Alternatively, it is possible to determine the threshold for detecting lights of different colours. In this way the sensitivity of the bird across the spectrum, its spectral sensitivity, can be determined. If lights of different colours are paired it is possible to determine how close they can be before they can no longer be told apart. This will give an indication of how fine the colour vision of the bird is. It is also possible to introduce flickering lights so that the slowest and fastest flicker speeds can be detected. This can have useful applications in that it gives a clue to how slow or fast a flickering light might need to be to act as a warning or distraction.

With other ingenuity and interests it can be seen how this kind of psycho-physical approach can be used with sounds rather than light. In these kinds of studies birds are not trained to respond when they see something but when they hear it. Systematic alterations of sound levels and frequency can provide insights into a bird's absolute sensitivity to sound, its relative sensitivity to sound of different frequencies (audiograms), and how finely the birds can discriminate between sounds of different frequencies. It would also be possible to introduce sounds with different time patterns, trills etc., and investigations with these can give insight into the ability of birds to detect differences in bird songs.

Some investigators have even been able to modify the overall procedure to work with smell and taste to determine which chemical stimuli are more salient and can be detected at low concentrations. However, controlling the smell or taste compounds and presenting them in a uniform way can be very difficult, and perhaps only a few trials can be done each session.

Who, how, and what to measure?

There are many ways in which such training techniques can be used to inves-tigate the senses of birds in a robust way. These allow comparisons between species. However, not all species will be readily amenable to such training, or investigators may not be able to invest the time necessary for the use of these training procedures in full. Some investigations have been able to use the basic principles of these training procedures to gain insights into birds' senses albeit in a rather limited way compared with the full descriptions that might be aspired to. For example, the first investigations of whether birds could detect ultraviolet light was done using a modified procedure with hummingbirds, and some early work on the sense of smell in doves was carried out using modifications of this kind of approach.

Much depends on the time available to do the work and the motivation of the investigators to work with their species. Unfortunately, modern ways of funding science and pressure to produce papers often preclude long-term training projects of the kinds described here, but work is still being done. For example, recently there have been some valuable investigations of vision in diurnal birds of prey and in parrots, using just these kinds of procedures. Valuable insights into what diving birds might see underwater have come from training investigations of these kinds. All of these studies took many months and required the accumulation of many thousands of trials with the birds.

There remain plenty of opportunities to pose questions about the sensory capacities of most bird species. As the comparative database of species' sensory thresholds grow, the results become of ever greater value.

Chapter 3

Vision in birds: the basics

Vision first emerged on Earth about 540 million years ago. The first eyes were simple structures able only to register changes in the amount of light falling upon them. However, these simple structures, and the limited information they provided, had a profound influence on the evolution of animals. From a modern perspective these first eyes were game changers, the equivalent of today's disruptive technologies. It is no exaggeration to suggest that the emergence of eyes changed forever the type, quality and quantity of information that underlies the interactions of animals with their environments, which, of course, includes their interactions with other animals. Ever since the simplest eyes evolved, the gaining of information about the world, its interpretation, and the uses to which it is put, has become increasingly complex and subtle.

The evolution of vision

Prior to the evolution of eyes, the ways that animals were able to gain information about their worlds was both slow and intimate. It depended primarily upon the transmission of chemicals through air or water. This information either took some time to arrive or it concerned objects that were very close to or, more likely, in direct contact with an animal's body. The key advantage that even simple eyes brought was that information about events remote from the animals were received immediately. This advantage was so dramatic that the first simple light-sensitive structures, which could detect just the presence and intensity of light, rapidly evolved into something far more sophisticated.

This was the birth of 'spatial vision'. This is the ability to determine not just that light is present and changing in intensity, but also the direction from which light is coming. The refinements of eyes and vision over the past 500 million years have been concerned primarily with elaborating the accuracy of spatial vision. This elaboration has been concerned with extending the degree of detail that can be extracted about light sources at various distances, extending the range of light levels over which this can be achieved, and increasing the volume of space about an animal from which information can be obtained at any instant.

The high utility of such information is indicated by the rapidity with which eyes evolved. It took probably less than 2 million years for eyes that simply registered the presence of light to evolve into 'camera eyes'. These are eyes that show all of the main features of the eyes that we recognise in species of the present day, including in ourselves. It has been argued that the evolution of sophisticated eyes, showing all of the key features of modern eyes, could have involved only about 400,000 generations of change (Figure 3.1).

The newly evolved ability of spatial vision provided information not only about objects that were close by, but also about those that were far away from the animal. By so doing it established vision as the primary source of information used to guide

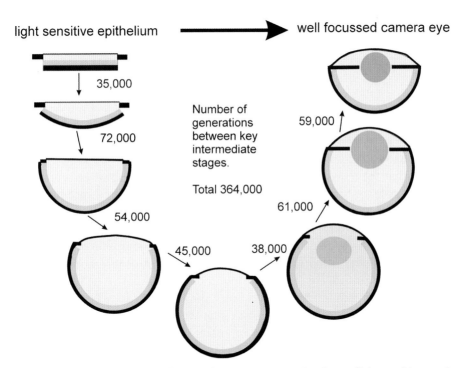

light sensitive epithelium ⟶ well focussed camera eye

35,000

72,000

Number of generations between key intermediate stages.

59,000

Total 364,000

61,000

54,000

45,000   38,000

FIGURE 3.1 The evolution of well-focused camera eyes, starting from a light-sensitive patch on the surface of an animal. The diagram shows a theoretical model based on conservative assumptions about selection pressure and the amount of variation in natural populations. This model, proposed by Nilsson and Pelger from the University of Lund, suggests that an eye could have evolved very fast, in fewer than 400,000 generations. The starting point is a flat piece of light-sensitive skin (shown in blue) with a transparent protective layer over it, and below the receptor cells a layer of pigmented cells (shown in black). These absorb light not caught by the receptors and help provide integrity of the whole structure. The emerging chamber is filled with a clear fluid and eventually by a lens, while the original protective layer becomes curved and eventually takes on an optical role as the cornea. This evolutionary pathway is not just theoretical, it is informed by the fossil record. (Redrawn from the original scheme proposed by Nilsson and Pelger in 1994.)

behaviour. This primary reliance upon vision is found today in nearly all animal taxa including, of course, birds.

Camera eyes (their basic design and functional divisions are described below) had been around for over 350 million years before the first birds appeared on the planet, about 150 million years ago. This means that the first birds, and their dinosaur ancestors, were highly likely to have had elaborate visual capacities. These capacities had been honed through the process of natural selection in response to the challenges of extracting information from the many different environments that had occurred on the earth over a number of geological eras. As sophisticated as these first bird eyes might have been, their evolution has continued. Changes in vision have occurred in response to the new environmental challenges that emerged as bird lineages diversified to exploit the wide range of habitats in which birds now exist. Across today's 11,000 bird species, eyes exhibit both major and subtle differences in design and function.

It is difficult to know the paths and time courses over which eyes have evolved. However, Dan Nilsson of Lund University has argued that the evolution of eyes has not been linearly progressive. It seems that eyes, like many other structures, have evolved in fits and starts. This means that relatively long periods of stasis in eye structure were punctuated by periods of rapid change, triggered by the emergence of new tasks as the environment changed. These changes will have included the availability of new food sources and feeding opportunities, and the appearance of new species that provided new threats and opportunities.

Just what these tasks were is very difficult to determine. However, the overall tasks that seem to have driven the evolution of eyes in birds are concerned with the control of the bill or feet towards specific targets, especially in foraging, and in the detection of predators. These two main sorts of tasks typically make substantial, but often conflicting, demands upon eye design and most aspects of bird vision. It seems likely that other tasks, especially the control of locomotion (flight, swimming, or walking) are achieved within constraints imposed by these two key tasks of foraging and predator detection. Surprisingly, it seems that the control of locomotion may not be a prime driver of vision in birds. These arguments will be expanded upon later, but the important point to note here is that vision, alongside other aspects of an animal's biology, is driven by its utility. In the case of vision (and other senses) that utility lies in gaining information for the control of behaviour. Vision of a certain kind is not just something that a bird happens to have, it must fulfil important functions.

The importance of vision in birds

The primary reliance of birds on vision is easily asserted. It can be based just upon casual observations of birds completing their everyday behaviours. However, this assertion is also well supported by evidence that in most species of birds relatively large portions of their brains are devoted to the analysis of information from vision.

Also, the so-called 'intelligent' behaviours of birds seem to be based primarily upon visual information. Thus, extracting information from vision and using it to guide sophisticated behaviours is the essential function of the brains of most birds.

Only in a handful of extant bird species is vision not the primary sense. Even in such birds vision was at one time highly likely to have been the prime source of information. However, vision can become secondary through a process of regressive evolution. At the same time other senses, particularly olfaction, hearing, and touch sensitivity, can come to take on some of the primary functions usually carried out by vision. These include food detection, predator detection, and guidance of locomotion – all of which are underpinned by vision in most birds.

The prime examples of 'non-visual' birds are the five species of flightless kiwi, which probably lost their reliance on vision as they evolved in the absence of mammalian predators on the islands of New Zealand. The downgrading, but not complete loss, of vision in these birds is discussed in Chapter 9.

In other bird species, including some of the petrels (Procellariidae), shorebirds (Scolopacidae), and owls (Strigidae), olfaction, touch, and hearing respectively play a key role or one that is complementary to vision, especially when it comes to finding and ingesting food items. However, in these birds vision is still the primary guide for locomotion.

What eyes do

The crucial property of the first camera eyes, and indeed of all eyes since, is that they were able to determine the position of a light source relative to the animal. They were more than simple light detectors; they also had the capacity of spatial vision. Spatial vision provides information on where objects are relative to the observer. What's more, it can do this more or less instantaneously and continuously.

This might seem an obvious attribute of vision, but it is not true of any other sense, nor was it an attribute of the very first eyes. Not until camera eyes evolved was it possible to obtain accurate information about the positions of objects within a large part of the environment in which an animal sits. Furthermore, most camera eyes can do this over a range of light levels, although as light levels fall the accuracy of spatial vision usually decreases. Being able to function over a range of light levels is an important attribute of vision because in natural environments light levels change both constantly and dramatically. In open habitats at all latitudes except close to the poles, ambient light levels change over many million-folds on a daily cycle; from noontide sunlight to starlight. Therefore, a key aspect of an animal's eye is not only how much spatial detail it can detect but also over how much of the daily light cycle it can provide useful spatial information.

Functioning over the full range of naturally occurring light levels is difficult. Some eyes have evolved to provide spatial information over a wide range of light levels, but many have evolved to function primarily within a relatively narrow range, typically those experienced during daytime (dawn to dusk) or night-time (dusk to dawn). Even within these periods light levels are not static and change over many thousand-folds.

Colour vision is primarily an elaboration of spatial vision. Colour vision is often thought of as something rather different or special, something that is additional to 'simple' spatial vision – perhaps regarded as simple because it can be achieved in what appears to be a less sophisticated world of black and white. However, colour vision has value because it enhances the extraction of spatial detail by using differences in how light of different wavelengths is reflected from different surfaces.

Lit by sunlight, a 'blue' surface reflects light only within a relatively narrow range of the wavelengths of light that fall upon it. The surface absorbs light from the rest of the spectrum. A 'red' surface reflects and absorbs light in other parts of the spectrum. However, the colour vision mechanism that determines which part of the spectrum light is from is rather wasteful of photons. This is because the visual system must make elaborate comparisons between light reflected from the different surfaces. The consequence is that the ability to detect levels of contrast between patterns is always lower for coloured than for black and white patterns. Faced with the task of detecting contrast in a grating (of the kind discussed above in the 'Measuring senses' section of Chapter 2), or with the task of resolving the finest stripe width that can be reliably detected, performance with stripes of different colours is always inferior compared with black and white patterns.

Lack of colour vision at low, night-time, light levels occurs in most vertebrates. It is not because there is no colour information potentially available in the environment. The lack of colour is a property of the visual system, not of the environment. At night, photons are relatively scarce. To see something at night necessitates maximum use of any photons that are available. Having a mechanism that detects the part of the spectrum that photons come from is too wasteful of light to have general utility compared with the advantage of simply being able to detect that something is actually present. The stimuli for colour vision are present in the environment at night as much as they are during the day, but vision does not make use of them. Colour vision is a bonus of high light levels.

This simple observation tells us that colour is not a property of the world but a property of the visual systems that extract information from the world. Light itself is not coloured. Colour is an attribute added by visual systems. This observation was first made by Isaac Newton in his *Opticks* (published in 1704) and captured in the famous phrase 'The Rays, to speak properly, are not coloured'. It was based initially upon his observations of how white light can be broken up into prismatic colours. Newton elaborated this key idea by further experiments on many aspects of human vision. The implications of this observation have been investigated and discussed from both scientific and philosophical viewpoints ever since. Humans have projected many aesthetic properties onto 'colour', and this has given philosophers a rich theme for speculation and theorising. It is essential, however, to be aware that colour vision is an elaboration of the mechanisms that extract spatial detail from the environment.

Sources of variation in vision

We are familiar with the idea that there are many different designs of bird wings and bills. Ornithology textbooks show arrays of different wing and bill types along with discussion of their different attributes and functions. Such discussions make it clear that there is not a single optimal wing or bill type. A single wing type cannot fly at all velocities, or support birds of different weights, or carry out all kinds of manoeuvres. A single bill cannot be an all-purpose tool for extracting and handling many types of food. What is optimal depends upon the task. This also applies to the senses of birds, especially vision.

It is relatively easy to understand the structural bases for the different types of performance of birds' wings or bills. For example, there may be obvious differences in the relative lengths and flexibility of bones, or in the number, hardness, and relative lengths of feathers. Bills also differ in their length, shape, and flexibility. It is not so immediately obvious when looking at an eye how variations in vision can arise. How is it possible for one bird to have quite different visual capacities from another?

Camera eyes

The camera type of eye is found in all vertebrates and in some invertebrates (octopuses, squids). A camera eye is also referred to as a 'simple' eye, and this label is not without good reason. Compared with the complexity of the multiple repeated structures which are found in the compound eyes of most inverte-brates, camera eyes are structurally and conceptually simple. The important point, however, is that within this simplicity of basic design there is great potential for variation in each of the key components. Both gross and subtle variations in these components can profoundly alter the vision of an animal, and hence change the information that different eyes can extract from the same scene.

The basic structure of a camera eye has just two key functional components, an image-producing system and an image-analysing system (Figure 3.2). Not only can these components show much variation, they can also vary in their characteristics independently of each other. The image in the eye of one species will be different to that of another, as will the ways that these images are analysed. Furthermore, with two eyes in an animal's head, they can be placed in different positions with respect to each other in the skull. This alters the region about the head from which visual information can be retrieved at any one instant and can profoundly influence what an animal can detect in the world that surrounds it.

These two main functional components of camera eyes are conceptually simple. The optical system produces an image of the world outside the eye, and the analysis system extracts information from that image. These two functional components can be matched in a straightforward manner to the main anatomical parts of an eye (Figure 3.2). Indeed, they can be matched to the key components

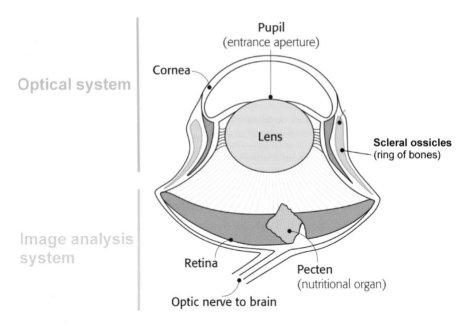

FIGURE 3.2 Camera eyes can be divided into two main functional parts; the optical system and the image analysis system. The optical system in bird eyes have two components, the cornea and the lens. These produce a focused image that is projected onto the retina, which is where the first stage of image analysis begins, and from where information is sent via the optic nerve to the brain. The key thing is that although these two functional components are joined together in a single eye, they can to some degree evolve independently of each other. The lens and cornea can have many different optical properties depending on their shapes and sizes and can produce images with different properties (for example, size, brightness, and contrast). The retina can exhibit huge variation in the way receptors are arrayed across its surface. This means that eyes with different image-making properties can evolve and eyes with different image analysis systems can also evolve. Even eyes which are of the same size and overall shape can have very different properties. Analysis of different eyes has revealed a plethora of subtle differences in both image production and image analysis. This results in the eyes of different species gathering different information about the world in which they sit. This diagram is based upon the tubular-shaped eyes found in owls. (Diagram by Nigel Hawtin, nigelhawtin.com.)

of any camera, from the camera in your phone to a sophisticated video, single-lens reflex, or plate camera.

The optical system of a camera eye consists of the lens and the cornea. The initial extraction of visual information is carried out by the retina onto which the optical system projects an image of the world. The retina is a very thin structure of immense complexity, made up of layers of specialised neural cells. These include the layer containing the photoreceptor cells, and it these which detect the pattern of light within the image. The neural cells of the retina are anatomically part of the brain to which each eye is connected via the optic nerve.

Although the retina shows immense complexity – indeed, it is composed of many millions of cells – there is only a small number of different cell types. This applies especially to the photoreceptors, whose types are discussed in more detail later in this chapter. The important point to note is that significant differences in vision arise from the ways in which the different photoreceptor types are packed together and arranged across the retinas of different species.

This variation in packing and arranging high numbers of receptor cells of just a few types is not unique to vision. It is what underpins variation in other senses too. This will be discussed in later chapters showing, for example, how variation in touch sensitivity, taste, and smell arise. Each of these senses is based on a relatively small number of receptor types, but marked differences in sensory capacity occur because of their relative numbers, and how the receptors are arranged, in different species. Ultimately, these variations are what underpin the sensory ecology of different species.

Sources of variation in camera eyes

These functional components of an eye can be matched to the two main functional parts found in all human-made imaging systems. From large astronomical telescopes to the small cameras built into mobile phones, these systems all have one part that produces the image and another that analyses it, and it is clear that the properties of these two components differ greatly.

Even within the cameras of mobile phones properties can be varied to give images that differ markedly in the information they provide. These differences in information capacities result primarily from three fundamental attributes: the degree of detail that can be detected, the extent of the world that is available for analysis, and the range of light levels over which the camera will operate. These will be familiar and important to keen photographers, but even manufacturers of mobile phones draw attention to these features in their marketing materials.

Comparing a mobile phone camera with an astronomical telescope is straightforward. Both are doing essentially the same thing in the same way, but the levels of information they provide are phenomenally different. However, neither one can do the other's job. The essential point is that the same consideration applies to eyes. They have evolved in different species to provide information for the conduct of different tasks and in different environments. Differences in their eyes are the result of relatively fine-tuning of both image production and image analysis, similar to the fine-tuning of components that underlies differences in phone cameras.

That people are willing to invest time and money in choosing between phone cameras indicates how differences in the information extracted by cameras are functionally significant. That eyes can differ markedly in all of these attributes suggests that if we were able to choose between different types of eyes, rather than having those we are born with, we might spend a lot of time in coming to a decision.

The optical systems of camera eyes

The optical system is composed of two main elements (Figure 3.2). The cornea is the relatively simple curved surface at the front of the eye. In eyes that operate primarily in air, the cornea is essentially a boundary between air and the fluid-filled chamber of the eye. The radius of curvature of the cornea is the key to its image-forming properties. A more highly curved surface produces a smaller image than a shallowly curved surface. The lens is suspended in the fluids that fill the chambers of the eye. It is also relatively simple. Like a magnifying glass, it has two convex surfaces, but these can vary in how curved they are. Unlike a magnifying glass, the interior of the lens is not uniform but is made up of a complex structure of transparent layers of different densities. The optical function of the lens is primarily concerned with making relatively fine adjustments that focus the image, already formed by the cornea, onto the retina.

It can be seen immediately that there is much scope for changing the overall image-forming properties of an eye by virtue of small changes in the absolute size, curvatures, and relative positions of these two optical components.

Although we cannot know the optical properties of the very first, relatively crude, camera eyes it is easy to understand that they must have varied in their properties with respect to a number of parameters. Two parameters were key, the brightness of the image (how much light is captured to make the image) and the quality of the image. The precision with which light from a point in the world is brought to a focus in the image determines how faithfully the image reproduces the world that it represents. Surprisingly small variations in optical structure result in marked differences in the way optics represent the world, and in eyes these small optical variations have been rich sources for the operation of natural selection. Selection for subtle differences in optics was the beginning of the process by which eyes have evolved to match the demands of different tasks and different light environments. Today we can identify eyes with marked differences in the brightness and quality of their optical images. Some examples will be discussed in later chapters that look at specific examples of the sensory ecology of birds which face different perceptual challenges.

Variation in image properties

By definition an image is never perfect. It is a simulacrum that always lacks some information about the world. The quality of the image, and hence the information that it contains, usually varies across the image surface. Image quality is usually closer to perfection along, or close to, the optic axis of the lens system. This is the line about which the optical elements of the system are arranged; in camera eyes it is the direction about which the cornea and lens are symmetrically aligned (Figure 3.2).

Moving away from the optic axis results in an image of progressively poorer quality. It is here, in the more peripheral parts of the image, where obvious distortions and aberrations occur. This is something that is readily apparent in simple hand lenses or in camera lenses at the cheaper end of the market. To correct for these peripheral distortions requires elaborations and refinements of the optical system, hence the high prices asked for camera lenses which maintain high quality across a broad section of the image.

The image produced by peripheral optics is often masked out in human-made cameras, and not presented for analysis by the film or photodiode array. However, peripheral optics cannot be ignored when trying to understand the visually guided behaviour of many vertebrate animals, including most birds. This is because in these species the eyes are placed on the side of the head and the visual field is often maximised, which requires use to be made of the entire image (Figure 3.3).

In some species, full use is made of an image from each eye that is more than 180 degrees wide. This gives the birds maximum visual coverage of the space around them. In many birds the width of the visual field behind the head is maximised in order to enhance the chances of detecting a predator, but this again can be achieved only by using peripheral optics (Figure 3.4).

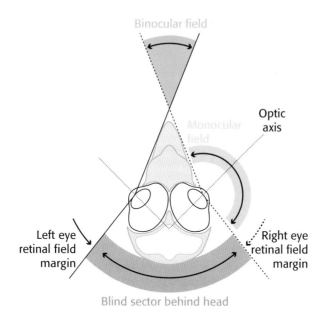

FIGURE 3.3 A diagrammatic section through the head of a bird showing a typical arrangement of eyes in the skull and how the visual fields of each eye combine. In all birds the eyes project laterally so that the axes of the eyes always diverge; no birds have forward-facing eyes. The fields of view of the two eyes are combined to give the total field of view, with a sector in front of the head where the two fields overlap to give a binocular field. A wide degree of variation in these basic arrangements is found in birds, resulting in different degrees of overlap, different width blind areas behind the head (though some birds have no such blind area). Just small variations in the width of the field of view of each eye, and of eye position in the skull, can result in large differences in visual fields between species. (Diagram by Nigel Hawtin, nigelhawtin.com.)

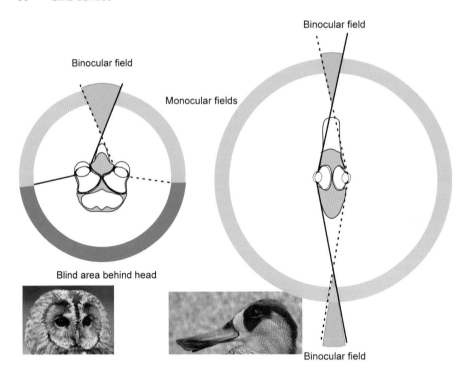

FIGURE 3.4 Examples of the extremes of visual fields found in birds. In a Tawny Owl *Strix aluco* the axes of the eyes project laterally and forwards. The field of view of each eye is relatively narrow, and the eyes sit in the skull to give a relatively large degree of binocular overlap and an extensive blind region behind the head. In a Pink-eared Duck *Malacorhynchus membranaceus* the fields of each eye are extensive, a little over 180 degrees, and they project laterally to give the bird a small degree of binocular overlap both in front of and behind the head. This means that it sees all around its head in the horizontal plane. In fact the binocular region extends right above the head, and the duck has panoramic vision of the hemisphere around and above its head.

This lateral placement of the eyes in the skull is quite unlike the situation in ourselves. The optic axes of our eyes, and hence the best-quality optics, project directly forward. We do not try to look in the direction that we are travelling out of the sides of our eyes. This is, however, what all birds do to some extent. No bird species, not even owls, have eyes positioned to face directly forwards, and many birds look forward with the very periphery of their eyes' optical systems. The consequence of this arrangement is that the best-quality optics in all birds projects laterally, away from the axis of the head, in some species markedly so. This has important consequences for understanding both the foraging behaviour and the role of vision in the control of locomotion in birds. It will be discussed in Chapter 5.

Another important property of the image is how much of the world is imaged at one instant. Does the imaging device have a wide or narrow field of view? This is important, since it determines from how much of the world around an animal's head information can be gained at any instant.

The image analysis system

It is the retina that starts the process of image analysis. The retina extracts the essential information that the image contains, encodes it neurologically, and sends it via the optic nerve for further analysis by the brain.

When looking into any eye we see the pupil as a black void. We are looking through the optical system and through the thin transparent neurological layers of the retina to the uniformly black surface, the pigmented epithelium, that lies behind it. As outlined above, each retina contains many millions of individual neural cells arranged in distinct layers. The most prominent of these layers contains the photoreceptor cells, the well-known rods and cones, and these are discussed in detail below.

When light photons reach a photoreceptor, a neural signal is generated and relayed to a ganglion cell which in turn relays that information through the optic nerve to the brain. Of crucial importance at this first level of image analysis are the number, density, and distribution of photoreceptor and ganglion cells across a retina. The actual numbers of photoreceptors are very high. For example, in the eyes of eagles (and humans) the total number of cells in the whole retina probably exceeds 100 million. In all retinas, however, the number and density of photoreceptor cells are far from uniform. In some locations photoreceptor cells are packed close together, in others they are more sparse, and large differences in density can occur between locations less than 1 mm apart. In an eagle's retina density can peak at about 450,000 photoreceptors per square millimetre, and in humans it reaches about 200,000, but less than 1 mm from the site of peak receptor density it drops to 16,000 photoreceptors per square millimetre. However, these changes in receptor density do not occur randomly across the retina; they occur in distinct patterns in the eyes of different species.

The patterns of photoreceptors can be revealed and characterised using isodensity contour maps (Figure 3.5). These link locations across a retina which have the same cell densities. In the same way that contour maps link locations of the same elevation and allow us to quickly appreciate the topography of a landscape, these density maps of retinal cells provide a quick way to compare retinas in different species and hence are a ready means of comparing some basic aspects of vision between species. A number of examples of receptor and ganglion cell maps will be discussed in later chapters.

The importance of these density patterns can be appreciated by considering the photoreceptors of a retina to be analogous with the photodiodes of the receptor surface of a digital camera. In a camera we understand that the photodiodes are responsible for pixelating the image, and we expect that the photodiodes are not jumbled but spread at an even density across the whole image-analysing surface. This guarantees that the same amount of detail is available across the whole of the image. However, in retinas the densities of receptors and ganglion cells vary markedly across the image surface. Furthermore, there are consistent and different photoreceptor and ganglion cell patterns in the retinas of every bird species. It is as

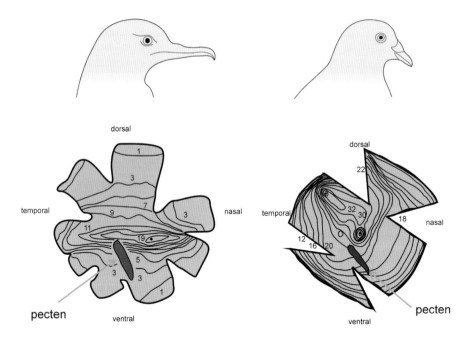

FIGURE 3.5 Examples of isodensity maps of the ganglion cells in bird retinas. In each diagram the retina has been spread flat, but its orientation is as in the eyes in the intact birds shown above. Flattening the retina causes splits, hence the rather ragged shape. The densities of the ganglion cells (×1000 per square millimetre) have been analysed across the whole of the retina and points with similar density have been joined to give contour maps, much as a topographical map links places of the same height. Clear patterns emerge in these maps showing how the images projected onto the retina by the eye's optics are analysed to extract different degrees of detail. On the left is the retina of a Manx Shearwater *Puffinus puffinus*, and the map shows that their retinas have a band running horizontally across the field of view in which details are particularly resolved (receptors are at high density, providing greater detail). On the right, the Rock Dove *Columba livia* shows a retina with two distinct areas from which detailed information is extracted: one looks out close to the axis of the eye (in this view almost directly out of the page), while the other also projects laterally but downwards and slightly forward within the bird's field of view. (The diagram of the dove is redrawn from work published by Bingelli and Paul.)

though we were able to choose between cameras not just on their overall density of photodiodes, but also on how the photodiodes are placed across the image surface. It is as if we could choose between one camera that analysed the image in greater detail at its centre, another that could analyse with greater detail in a band horizontally across the middle of the image, another more to one side, and so on. An endless number of possible arrangements would be possible.

Such patterns are indeed found in bird retinas, and indeed in all retinas, including our own. In human eyes the image is analysed in greatest detail more or less at its centre, and in less detail towards the periphery, but even in our eyes the patterns are not symmetrical. In birds highly complex patterns are found (Figure

3.5). What this means is that across bird species there is a wide range of patterns in the way that information is extracted from the environment surrounding the bird. The eyes of two different species may be imaging the same scene, but because of differences in their retinas the scene is analysed in different ways. These different receptor patterns are the result of natural selection driven by the need for the extraction of key information used in the control of different visually guided tasks in different species.

While knowing about these patterns gives valuable insight into the vision of different species, knowledge of the absolute density of ganglion cells or photo-receptors in a particular eye is also of great value. It allows an estimation to be made of the upper limits of the resolution (acuity) of an eye and can also provide an idea of acuity in different parts of the field of view. To achieve this, data on ganglion cell density must be combined with information on the size of the image. Using this method, estimates of the maximum spatial resolution of eyes have been calculated for a range of species (see Appendix). Furthermore, this method has been validated by determinations of maximum acuity using training techniques (of the kind described in Chapter 2) in the same species, most notably in some birds of prey, including both eagles and falcons.

Photoreceptors: rods and cones

Photoreceptors are of two basic types, rods and cones. The rods have higher absolute sensitivity and function principally at low light levels (between dusk and dawn in natural environments) while the cones underpin vision during higher, daytime light levels. Cones are the photoreceptors that provide colour vision.

Every photoreceptor contains millions of photosensitive molecules (visual pigment), and each molecule is capable of trapping an individual photon of light. Trapping just a small number of photons is sufficient to kick off the process of phototransduction. This is a cascade of chemical changes which results in a neural signal being generated and transmitted to the brain via the optic nerve.

An important aspect of visual pigment molecules is that they do not absorb light of all wavelengths. Each molecule is selective within the spectrum, responding maximally to a relatively narrow range of wavelengths. Because only one type of photopigment occurs in each photoreceptor they sample the spectrum of the light falling upon them. The result is that for a particular eye the relative numbers of receptors with different photopigments will determine its sensitivity across the spectrum. This is explored in more detail in the 'Types of cone photoreceptors' section below and in Box 3.1.

Sensitivity in the spectrum

It is clear that many birds detect light over a wider range of wavelengths than humans are able to; in other words, they have a broader visible spectrum. Also, it seems that birds may be able to discern more colours within their spectrum;

that is, they can probably make finer colour discriminations, at least in some parts of the spectrum. Some birds can detect light in the ultraviolet (UV) part of the spectrum, light to which human vision is insensitive. However, vision in the UV part of the spectrum is not unique to birds, for some terrestrial mammals and many invertebrates are also able to detect information using UV light.

It is important to note that not all birds see in the UV part of the spectrum. In fact those bird species which have true UV vision are found only in the gulls (Laridae, Charadriiformes), ostriches (Struthioniformes), parrots (Psittaciformes), and the oscine passerines (Passeriformes), but excluding the crows (Corvidae). Other bird species may have visual sensitivity that extends into the violet spectrum, but they cannot be considered truly UV-sensitive, while others, notably some birds of prey, have optical systems that filter out UV light from the image so that it never reaches the retina.

Colour vision

As noted in the 'What eyes do' section above, the prime function of colour vision is that it allows the extraction of spatial detail by using differences in the wavelengths of light that make up the image. It seems unlikely that colour vision evolved specifically for enhancing information about the presence and properties of certain types of objects that are key in the life of an animal. It is more likely that it first evolved to meet a broad range of spatial tasks.

Examples of situations where detection is enhanced by colour vision in birds include the use of objects in display behaviours (as seen, for example, among bowerbirds (Ptilonorhynchidae, order Passeriformes)), and the detection of plumage patterns and particular flowers or fruits against foliage backgrounds. This does not mean, however, that these specific tasks were the prime drivers for the evolution of colour vision. Indeed, it seems more likely that the colour of plumages, flowers or fruits evolved to become more conspicuous in response to the vision of the observing birds, rather than that the vision of the birds evolved to detect these particular objects.

Fruits that are eaten regularly by birds have probably evolved to encourage their detectability. This is because consumption of the fruit will result in wider dispersal of seeds. Thus, many fruits have evolved to be detected by birds and this is achieved in two ways. First, the ripe fruits (ready for seed dispersal) reflect light that contrasts with the surrounding foliage, both in brightness and in colour. Second, smaller fruits often occur in conspicuous concentrations, so that they present a larger target that can be detected with relatively low-resolution vision at a distance.

Figure 3.6 shows some examples of fruits whose seed dispersal is aided by birds. It is clear that they reflect light from different parts of the spectrum, from the red and orange wavelengths to the blues and ultraviolets. Interestingly, the same bird species may consume fruits with all of these different colours, suggesting that there

FIGURE 3.6 Fruits that are commonly eaten by birds in the British Isles, all of which have been recorded in the diets of European Robins *Erithacus rubecula*. Top row, left to right: Common Buckthorn *Rhamnus cathartica*, Elder *Sambucus nigra*. Middle row: Blackthorn (Sloe) *Prunus spinosa*, Common Juniper *Juniperus communis*, Spindle *Euonymus europaeus*. Bottom row: Hawthorn *Crataegus monogyna*, Rowan *Sorbus aucuparia*.

are no specialisations of vision associated with detecting fruits of particular colours. For example, European Robins are known to consume all of the fruits shown in Figure 3.6, plus many more.

All of these fruits occur in large concentrations, creating a large target for the bird to detect. This suggests that their detection does not require vision of high spatial resolution. As Figure 3.6 shows, although the individual fruits tend to be relatively small, they usually occur in large clusters. Therefore, the initial task for a bird is to detect these clusters at a long distance. Once it has found a cluster it is more or less guaranteed access to a large food source. Individual fruits can be detected when the birds work over the tree or shrub, but they can find this food source at a distance because of the large aggregations of individual fruits.

The ability to detect colour differences within the spectrum has not been determined directly in many bird species. Although colour might seem conspicuous, it is surprisingly difficult to show definitively that an animal is using colour vision. This is because surfaces may be discriminated both by colour and by brightness at the same time. To sort between these possibilities requires careful, and typically laborious, behavioural experiments. The most detailed knowledge available is from behavioural and electrophysiological studies in Rock Doves (Feral Pigeons) *Columba livia*. It is, however, possible to say something about colour vision, and the breadth of the spectrum visible, in birds in general.

This is based on knowledge of the visual sensitivity of individual cone receptors, and it seems safe to assume that practically all birds have colour vision. Furthermore, the ability to differentiate colours in much of the spectrum differs little between species. Even the nocturnally active owls are thought to have some colour vision, although it is not as sophisticated or as subtle as that of other bird species. However, genome analysis, although not the determination of visual pigments directly, suggests that colour vision may be absent from the kiwi species, and these may be the only group of birds that are not capable of making colour discriminations.

Types of cone photoreceptors

Classifying the different types of cone photoreceptors in the retinas of all animals is done by reference to the position in the spectrum of the peak sensitivity of the photopigments that they contain (Box 3.1). In our own retinas there are rods that contain a single type of photopigment, but our cones are of three types defined by which one of three different photopigments they contain. These pigments, and hence the cone types in which they occur, are commonly labelled red, green, and blue, to indicate the spectral regions in which they are most sensitive. It is these three types of cone receptors that provide the basis for human colour vision. It is referred to as *trichromatic* vision, because it is based on three receptor types.

In birds, while there are rod receptors containing a single type of photopigment (which is very similar to that found in humans), the photopigments of the cone receptors are typically of four types, giving them *tetrachromatic* vision. Three of these photopigment classes show a high degree of similarity across a wide range of species, but the fourth type can occur in two forms. One has maximum sensitivity in the UV part of the spectrum while the other has maximum sensitivity at violet wavelengths but does have some sensitivity into the near UV. It is this photopigment that gives certain birds vision in the UV.

The basic uniformity in visual pigments and receptor types across bird species provides evidence that there has not been adaptive evolution of visual pigments among birds. It suggests that photopigments and colour vision arose early in bird ancestry. Indeed, they were probably inherited from dinosaur ancestors, and their properties have changed very little over at least 150 million years.

Box 3.1 The photoreceptors of bird retinas

At one level of analysis bird retinas are highly complex. Even the smallest eyes contain many millions of individual photoreceptors, the rods and cones. However, at another level there is relative simplicity. This is because these photoreceptors are of few types and they are very similar in all bird species. Their essential features are depicted in the accompanying diagram. What varies between species are the relative numbers and distributions of the different receptor types across their retinas. As discussed in the 'The image analysis system' section above, it is these patterns of receptor distribution that are the primary foundations of differences in the visual abilities of species.

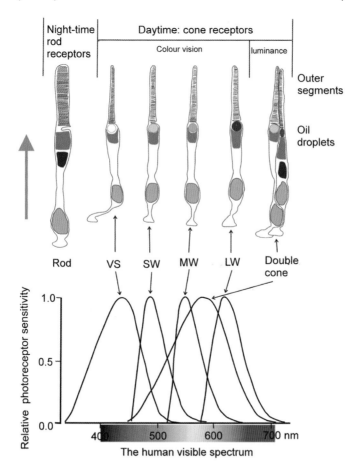

The cone receptors are the ones that function primarily at higher, daytime light levels. Rods function at low light levels. During twilight both rods and cones may function, depending on the exact light level. In birds, rods are of one type, but the cones are of five different types: four types of single cones, and double cones. The different types of single cones are classified by reference to the position in the spectrum of the peak sensitivity of the photopigments that they contain.

The top portion of the diagram depicts the types of retinal photoreceptors as they appear when viewed through a relatively high-powered light microscope. The outer segments are extremely narrow, generally between 1 and 2 μm (microns) in diameter, but they are relatively long. Each outer segment contains many millions of photosensitive pigment molecules, and each individual molecule of photopigment can absorb the energy from a single photon of light. When this happens, the receptor triggers a signal to the brain. Of course, at high light levels many millions of photons are simultaneously absorbed by the pigment molecules scattered throughout the outer segment.

The four types of single cone provide the fundamental mechanism upon which colour vision is based. The double cones provide a neural channel that is thought to signal luminance (brightness) and they are not part of the colour vision system. Within all cone types there is an oil droplet. As depicted in the diagram, most oil droplets are highly coloured. The colour is due to carotenoid pigments which are derived from the bird's diet. Birds fed on carotenoid-free diets have colourless oil droplets.

The blue arrow on the left indicates the direction in which light travels from the optics of the eye to the focused image on the retina. In one sense the retina would seem to be back to front, because light does not reach the photopigment until it has travelled through the neural layers of the retina. However, this arrangement means that the light that makes up the retinal image must pass through the oil droplets before it enters the outer segments.

The fact that the oil droplets are coloured means that they have light-filtering properties: they let light of certain wavelengths pass through and absorb light of other wavelengths. In fact, the oil droplets act as cut-off filters, that is they allow light only above particular wavelengths to pass through to the pigment molecules, while light of shorter wavelength is absorbed. The combination

A clear example of the uniformity of photopigments across modern birds comes from a study which showed that the visual pigments found in the eyes of a species of pelagic seabird (Wedge-tailed Shearwater *Ardenna pacifica* from the Procellariidae, order Procellariiformes) are very similar to those found in a phylogenetically distant species that lives in open forest habitats (Indian Peafowl *Pavo cristatus* from the Phasianidae, order Galliformes). This suggests that colour vision in birds has general, all-purpose, properties, that are not tuned to specific tasks performed by different species in different environments.

of photopigment types and oil droplet types results in there being four main types of single-cone photoreceptors in birds' retinas, with each cone type able to absorb light only within a particular part of the spectrum, although there is overlap between them.

The lower section of the diagram shows the resultant 'photoreceptor sensitivities' and the labels used to describe them. These are LW (long wave), which absorb light at the orange–red end of the visible spectrum; MW (middle wave), absorbing light in the green–yellow spectral region; SW (short wave), absorbing light in the blue–green spectral region; and VS (violet-sensitive), which absorbs light in the violet-ultraviolet spectral region.

The photoreceptor pigments in the LW, MW, and SW types of cone receptor differ one from another, but each cone type is highly similar across all birds species. However, the VS cones can contain two different types of photopigments. One type has a peak sensitivity in the violet at about 410 nm, which is within the human visible spectrum. The other type has its sensitivity centred around 360 nm, which is in the part of the spectrum not visible to humans. It is referred to as the UVS pigment. Cone photoreceptors that contain UVS pigment are found only in songbirds (oscine passerines) and in some non-passerine species: gulls, ostriches, and parrots. It is those species which have the UVS photopigments that have true ultraviolet vision.

Double cones are widespread in vertebrates but are absent from mammals. In birds, the double cones always contain one type of pigment. It has broad sensitivity across the spectrum centred on about 565 nm in the yellow region of the visible spectrum. The rod photoreceptors are found across the retinas of all vertebrate eyes and have a similar broad sensitivity centred about 500 nm, in the green part of the spectrum. Rods are considerably larger than cones and are usually found across the whole of the retina. However, they are usually absent from the fovea, which is where cones are found at their highest concentration. In species which have the highest acuity, such as raptors, double cones are also absent from their foveas.

Colour through birds' eyes

Just how the world might look when viewed through the birds' tetrachromatic colour vision system is of considerable interest. For example, does tetrachromatic vision mean that particular parts of the spectrum have different salience to birds compared with our own view of the world? Certainly, knowledge of the tetrachromatic system suggests that birds'-eye views are likely to be different to how humans detect colour information in the world. Of course, it is not possible to see these colours through birds' eyes, but computational methods have sought to compare colour patterns viewed through trichromatic and tetrachromatic vision systems, and have given some clues as to which kinds of differences between natural colour patterns are changed by a tetrachromatic system.

All of these differences in spectral sensitivity and colour vision, however, apply only at high (daytime) light levels when the cones are active. At lower (twilight and night-time) light levels, only the rods function and they have very similar characteristics across all birds and across mammals. Thus, at low light levels a similar sensitivity across the spectrum, and an absence of colour vision, is found in all bird species and indeed in humans and other mammals.

Chapter 4

Bird eyes: variations and consequences

The previous chapter made clear that there are many sources of variation in bird eyes involving both the optics and the retina. In this chapter specific examples of variations in eyes and their consequences for vision are described. The coverage of bird species is far from exhaustive. The eyes of relatively few species have been looked at in detail. However, even from these examples it is possible to grasp how much the vision of birds can vary, and how these variations can be interpreted as meeting the different sensory challenges posed by different tasks in different environments.

The overall message is that how birds see the world, that is how they extract information from the world about them, can differ markedly between species. Conspicuous differences in the bills, feet, and wings of birds are easily understood as variations in the tools that allow different species to exploit resources in different ways. Less conspicuous, but equally important, differences also exist in the information that different species have available to control those tools.

We can start to get an insight into these differences in vision by looking closely at some examples of variation in the optical structures that produce an image of the world. After that, some examples of differences in retinal structures will be considered.

Image production: optics of the eyes of doves and shearwaters

A good example of how differences in optical structure can influence vision in birds in a functionally relevant way is revealed by comparing the eyes of two familiar species, Rock Doves and Manx Shearwaters (Figure 4.1). These species have been chosen because their eyes are the same size, yet casual observations, even of birds held in the hand, give no clue to the structural differences in their optical systems. However, their optical systems do differ and result in important differences in the images formed in the eyes; differences that can be related to key aspects of their behaviour and ecology.

Rock Doves and Manx Shearwaters have quite different evolutionary origins, being from two well-separated avian orders, Columbiformes (pigeons and doves)

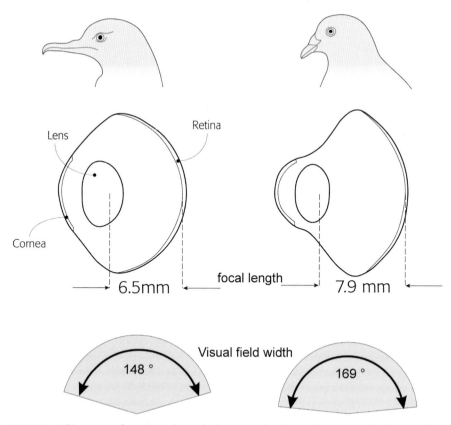

FIGURE 4.1 Diagrams of sections through the eyes of a Manx Shearwater *Puffinus puffinus* and a Rock Dove *Columba livia*. The eyes of these birds are of the same overall size (in both length and diameter), but it is clear that their corneas and lenses have quite different characteristics. The result of these differences in the optics is that the images produced by these eyes have quite different properties. The image is smaller in the shearwater (indicated by a shorter focal length) and brighter. Also, the visual field of a shearwater's eye is significantly narrower than in a dove.

and Procellariiformes (tube-nosed seabirds) respectively. They are markedly different in general structure, behaviour, and ecology.

Although the domesticated forms of Rock Doves, which are often referred to as Feral Pigeons, are mainly birds of towns, the wild ancestral populations nest and roost on cliffs, often above the sea. Indeed, the only UK populations of ancestral Rock Doves are thought to occur on sea cliffs in northern and western Scotland. Rock Doves are strictly diurnal, active only during the day when light levels are high. They feed on small seeds and leaves that are acquired by pecking.

The behaviour and ecology of Manx Shearwaters could not be more different. They are birds of the open oceans, coming ashore only to breed and then mainly under the cover of darkness. Their nests are usually placed in deep burrows in the ground above cliffs, or on steep slopes overlooking the sea. They feed, frequently at night, by seizing fish and other marine organisms from the surface, or during shallow dives.

The eyes of both Rock Doves and Manx Shearwaters are just under 12 mm long and they have a maximum diameter of about 14 mm (Figure 4.1). These dimensions are about half those of human eyes. The only external suggestion that their eyes are different is that the iris in a dove is usually conspicuously coloured whereas in shearwaters it is dark brown and difficult to see. However, this difference is superficial, since iris colour does not have a visual function and it cannot influence the properties of the image formed in the eye. Iris colour in birds is concerned with social behaviour, signalling such things as age, sex, emotional state, and general fitness.

Analysis of the optics of these two eyes reveals marked differences in the structure of their corneas and lenses, and in their relative positions. The result is that the shearwater's eye has a shorter focal length than the dove's and a wider entrance pupil. Thus, although both eyes produce a focused image on the retina, the parameters of those images are noticeably different. The retinal image in shearwaters is both smaller and brighter than in doves. When the bird looks at the same scene through the same-sized pupil, the image in a Manx Shearwater's eye is 1.5 times brighter than in a Rock Dove's eye. When pupils are fully dilated the shearwater's image is even brighter, perhaps twice as bright as the dove's.

This difference is functionally significant. It can be related to the nocturnal behaviour of shearwaters or, viewed from the other perspective, to the reluctance of doves to fly at night. These differences in image brightness are further amplified by the fact that in shearwaters the retina is rich in rod receptors and has fewer cone receptors, compared with the dove's retina. This means that shearwater eyes are better able to trap the photons that make up the dimmer images of night-time scenes. The smaller image in the shearwater's eye, however, means that there is less spatial information available in the image. Furthermore, the shearwater's image is spread over fewer photoreceptors and so the spatial resolution (acuity) of shearwaters will be lower than in doves.

A further important difference between these two eyes is in the fields of view. The angular width of the field of view determines how much of the world is represented in the image, and therefore how much of the world can be seen in a single glance. The image in the dove's eye is about 15% wider (the actual values are 169° and 148° respectively) (Figure 4.1). This is quite a large difference and could be behaviourally important. It means, for example, that at a distance of 100 m a dove's eye produces an image that covers a section of the world that is 295 m wide, while the shearwater's eye takes in a sector only 259 m wide. These differences are not trivial, and the wider field of view would enhance the probability of a dove being able to detect a potential predator.

Transferring these findings from eyes to photographic cameras, we would be advised to choose a camera based on a shearwater's eye to take photographs of dimly lit scenes. A camera based on a dove's eye would be better suited for daytime use, when it would give photographs with finer details and would also capture a wider part of the scene.

Image production: optics in the eyes of starlings, owls, and ostriches

The comparison of Rock Dove and Manx Shearwater eyes shows that there is not a 'standard design' of bird eye optics. It suggests that the optics of each eye is likely to be a specific solution to the challenges posed by different tasks and environments. The eyes of shearwaters and doves are examples of eyes that are of the same size – but what happens in species in which eyes differ markedly in size?

An indication of what can occur is shown by comparing the optics of eyes that are close to the extremes of the range of eye sizes found in birds. In this comparison, eyes that are among the largest of bird eyes, those of a Common Ostrich *Struthio camelus*, are compared with the eyes of a strictly nocturnal owl (Tawny Owl *Strix aluco*), and with the eyes of a small passerine that is active during daytime (Common Starling *Sturnus vulgaris*). The axial lengths of their eyes are approximately 38, 28, and 8 mm, respectively. The whole head of a Starling could fit comfortably within the eye of an Ostrich.

Ostrich eyes appear to be huge, perhaps because they conspicuously protrude from the skull. However, they are not uniquely large among birds. For example, eyes of similar size have been recorded in some species of albatrosses, penguins, and eagles. However, in these species the eyes are more enclosed within the skull and are not so conspicuous.

The eyes of owls do not sit within the skull at all; they also protrude and sit more or less on its surface (Figure 4.2). This is not obvious because owl eyes are surrounded by feathers. Owl eyes are held immovably in place by the same muscles that are used in most bird species to control the movements of eyes that are protected within sockets. Human eyes also sit deep in sockets, but they are smaller than those of Ostriches and Tawny Owls, with an axial length of about 24 mm.

The eyes of the owls, starlings, and ostriches are shown diagrammatically to scale in Figure 4.3. It is clear that these eyes have very different overall shapes and focal lengths. Owls' eyes are tubular with concave sides. This shape does not have an optical function but serves to reduce the volume, and hence weight, of the eyes. If owl eyes had a more flattened spherical shape (like those of ostriches and starlings) they would protrude even more from the skull than is shown in Figure 4.2. They would be even more difficult to hold in place and would require a much larger skull.

Differences in the focal lengths of these eyes are clearly evident (Figure 4.3), and they are important to know because they determine the size of the image available for analysis by the retina and ultimately the spatial resolution of the eye. The larger the image of a particular object the higher the potential resolution, and this is at the root of why photographers seek lenses with different focal lengths for capturing images of different objects.

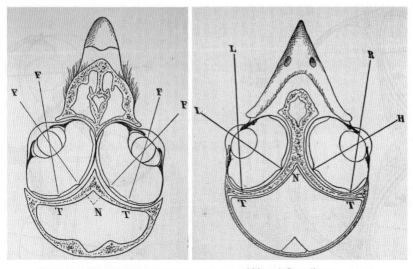

Broad-winged Hawk   Wood Swallow

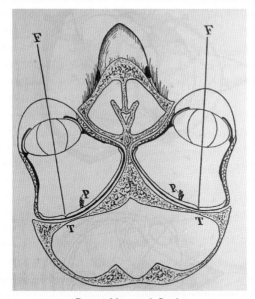

Great Horned Owl

FIGURE 4.2 Diagrams of horizontal sections through skulls of birds, showing how the eyes are usually enclosed snuggly within the protection of the skull, but in an owl the eyes are hardly within the skull. The result is that owl eyes are almost immovable and are simply held firmly in place by the muscles which in other birds control small movements of the eyes. All three of these examples show that the eyes take up a large volume within the skull and meet in the middle of the head. In some bird species the eyes are kept apart by a thin septum of bone, but in others the eyes touch each other. (These illustrations are taken from the book *The Fundus Oculi of Birds* by Casey Albert Wood, published in 1917: see the 'Image analysis: comparing retinas' section below.)

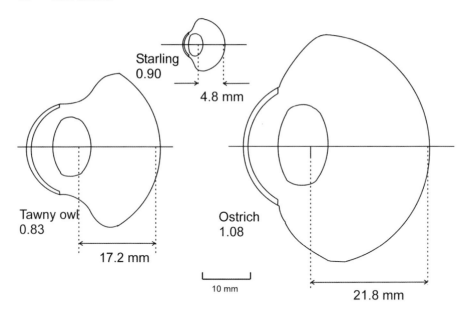

Starling
0.90

4.8 mm

Tawny owl
0.83

17.2 mm

Ostrich
1.08

10 mm

21.8 mm

FIGURE 4.3 Sectional diagrams of schematic models of the eyes and optics of Tawny Owl, Common Starling, and Ostrich. These diagrams are to scale and show the huge size variation found in the eyes of birds. Also indicated are the focal lengths of these eyes and the ratio of the power of the lens to the power of the cornea. The greater the focal length the larger the image of the same object when viewed by the eye. The ratio of lens to corneal power is another key descriptor of the optics of camera eyes and is explained in the text.

A useful way to differentiate between the optics of eyes that is independent of their size and focal length is to compare the relative contributions that the lens and the cornea make to the overall refractive power in each eye. This comparison indicates how much each of the two main optical components contributes to producing the image focused on the retina. This is captured by the ratio of the refractive power of the lens (F_L) to the refractive power of the cornea (F_C), the F_L:F_C ratio, and these ratios are also given in Figure 4.3 for each eye.

Refractive power is the common metric used for describing any optical component. For example, it is what an optometrist details when providing a prescription for spectacles, or it is what you might see on the label when selecting a pair of reading glasses. The units used are dioptres (D). This is the reciprocal of the focal length of the lens expressed in metres; the focal length is the distance from the lens at which the image of distant objects is brought to a focus. For example, a lens which creates an image at a distance of 100 mm (0.1 m) has a power of 1/0.1 = 10 dioptres, a lens with an image distance of 20 mm (0.02 m) has a power of 1/.02 = 50 dioptres, and so on.

Comparisons of the F_L:F_C ratio (Figure 4.3) show that the eyes of Ostriches, Tawny Owls, and Starlings are optically different, with ratios of 1.08, 0.83, and 0.90 respectively. If the ratios of Rock Dove and Manx Shearwater eyes (Figure 4.1) are also considered (0.4 and 1.6, respectively) it is clear that all of these bird eyes differ one from another not only in their absolute sizes but also in how they form an image of the world. Furthermore, each of these optical systems will produce images of different width and different maximum brightness.

Tawny Owls have a maximum image brightness twice that of the Starlings and Ostriches, and this presumably reflects the highly nocturnal behaviour of owls versus the strictly diurnal activity of Common Starlings. However, a Starling's eye gains a much broader view of the world – 161°, compared with the owl's 124° – and this can be seen to function in a Starling's need to be vigilant for predators, in its highly social nature, and in its ability to form dense flocks both on the ground and in flight.

It is perhaps surprising that the eyes of starlings, ostriches and owls, despite differing in size over a nearly five-fold range, are optically rather similar, while the dove and shearwater eyes (Figure 4.1) are very different, although they are of similar size. What these comparisons indicate is that the image-producing systems of birds' eyes can indeed differ markedly. There clearly is much flexibility inherent in the optical systems of camera eyes. What is found in each species is not random; it reflects the natural selection of different features which enhance information gathering when the bird is conducting different kinds of tasks.

Image analysis: comparing retinas

In Chapter 3 (Box 3.1) we saw that there are just six types of photoreceptors in bird retinas: rods, four types of single cones, and double cones. It was also seen that the photopigments these receptors contain vary very little between species. In evolutionary terms the cones and rods of bird retinas are 'highly conserved', meaning that they evolved very early and have been retained as species differentiated.

This small number of photoreceptor types is the basic component of the arrays of receptors found spread across the retinas in every bird species. These receptor types can be considered analogous to the small number of different types of photodiodes that analyse the image in a digital camera. Like a camera's detector surface, every retina contains many millions of photoreceptors. However, unlike in a camera, retinal photoreceptors are not distributed uniformly across the retina, nor are their patterns of distribution the same in different species. These differences in photoreceptor distributions are far from trivial, and it seems possible that the retinas of no two species of birds are the same.

Understanding how photoreceptors, and the ganglion cells to which they connect, are arranged in different species is of great importance for understanding the sensory ecology of birds.

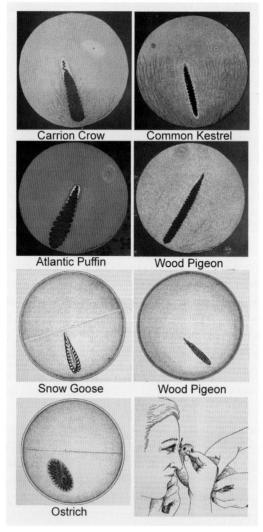

FIGURE 4.4 Examples of views of the retinas of different bird species as shown by Casey Albert Wood in *The Fundus Oculi of Birds* (1917). The technique was simple enough, and in the bottom right-hand corner is an illustration of it, also taken from the book. This involved just looking into the eye with a simple ophthalmoscope and carefully recording what was seen. These sample illustrations reveal a wide variety of different structures and colours in birds' retinas. Although the four coloured illustrations are difficult to interpret, they probably reflect marked differences in the distributions of rods and cones across the retinas. The lower illustrations emphasise the areas of the retina where there are heightened concentrations of photoreceptors. In the examples of Snow Goose *Anser caerulescens* and Ostrich a linear band running right across the retina where receptors are highly concentrated is clearly seen, while the Wood Pigeon *Columba palumbus* does not show any obvious regions where the concentration of receptors is increased. Note that, unlike mammals, birds have no blood vessels in the retina. Nutrition of the retina is provided by diffusion from the pecten. This is the large black structure which is shown in each of the illustrations. Therefore, the patterns of red seen in these illustration are not to do with blood vessels.

This is for three reasons. First, maximum resolution (acuity) can differ markedly between the eyes of different species as a result of differences in the packing density of photoreceptors. Second, in all eyes, resolution varies across the image, and highest resolution occurs in only a limited part of the field of view of an eye. Outside of those restricted areas spatial resolution can be a lot lower. Third, there are very marked differences in patterns of image analysis between species, and these will often involve differences in colour vision, not just resolution, in different parts of the field of view.

The consequence of these differences is profound since they suggest that analysis of the exact same scene by the retinas of different species will be quite different. The challenge for a sensory ecologist is to understand the reasons for those differences – not only how but why retinas differ between species.

Differences in image analysis within the field of view were first quantified in human eyes 300 years ago. However, it was only in the early part of the twentieth century that variation in image analysis in birds was recognised. The first detailed studies of variability within the retinas of birds was stimulated by Casey Albert Wood's comparative studies, published in his book *The Fundus Oculi of Birds* in 1917. Wood simply looked into the eyes of living birds with a hand lens or simple ophthalmoscope and carefully recorded what he saw of the retina. He also looked at the retinas of excised eyes under a low-power microscope.

Wood observed that there were marked differences between species in the colours of their retinas (Figure 4.4), and he was able to discern discrete areas of various shape in the retinas of some birds. He also showed that such features differed in their positions within the retinas of different species. In some bird retinas he described linear features, such as a band running roughly horizontally across the whole retina. These have become known as 'horizontal streaks' or 'linear areas'. These features were subsequently shown to be indications of regions where photoreceptors and ganglion cell densities are high compared with the rest of the retina. They are in fact slight swellings in the thickness of the retinal layers due to a higher density of cells.

Inspired by this early descriptive work, anatomists and microscopists developed sophisticated techniques to examine bird retinas, and many patterns of receptor and ganglion cell distributions have now been described and their characteristics quantified. Some of these patterns have been related to the behaviour and ecology of the species, but the functions of some patterns have yet to be explained. Many more species remain to be investigated.

Variations in the distributions of photoreceptors and ganglion cells

Two major ways have been identified in which image analysis can vary between the retinas of different bird species. The first is variation in the spatial distribution of cone receptor types. These variations can be marked, and it seems likely that they result in differences in colour vision and spectral sensitivity in different parts of the visual field of a single eye. The second is variation in the distribution of ganglion cells, which results in differences in spatial resolution within the visual field.

Distribution patterns of cone types

A most striking distribution pattern of cones types was described in Rock Doves by Yves Galifret in 1968 (Figure 4.5). In the dove's retina there is a large area dominated by photoreceptors containing red oil droplets, which are long wave (LW) sensitive cones (Chapter 3, Box 3.1). This area looks downwards within the visual field, while receptors containing yellow oil droplets (the middle wave (MW) sensitive cones) predominate in areas which look laterally and upwards. These different areas are so prominent that they have become known as the red and yellow fields. They are obvious even to the naked eye in an excised retina that is illuminated from behind; there is a clear boundary between them.

The visual ecology and function of this striking regional specialisation within dove eyes is, however, not understood with certainty. It has been suggested that the upward-looking yellow field in some way enhances the contrast of objects seen against the blue of the sky. However, this is argued by analogy with a general yellow filter placed across the whole of the image (an effect seen when humans wear yellow filters before their eyes) rather than considering the sensitivity of individual receptors and how they interact.

Although not so striking as these fields in Rock Dove retinas, other work has shown systematic differences in the distributions of cone receptor types (classified by the colour of their oil droplets) in the eyes of other birds. These tend to show a gradient in the relative abundance of different receptor types across the retina, without such a clear boundary as in the retinas of Rock Doves. These gradients suggest that there are marked interspecific differences in both colour vision and spectral sensitivity across the retinas of many birds.

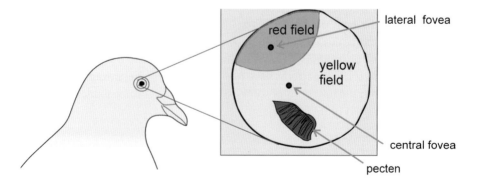

FIGURE 4.5 The red and yellow fields of the retinas of a Rock Dove. The diagram shows the retina of the right eye. It is depicted in a similar fashion to those shown in Figure 4.4 but in this case the retina is illuminated from behind and the light shining through picks out the areas where red and yellow oil droplets (see Box 3.1) are concentrated.

The functions of these differences are unclear. It seems likely that they are correlated with the perceptual challenges posed by life in different environments, and/or the conduct of different tasks. For example, it has been suggested that there is a generally higher proportion of cone receptors containing red droplets in birds that fly over water and need to see through the surface. In support of this idea are data from such diverse groups as kingfishers (Alcedinidae), gulls and terns (Laridae), and Northern Gannets *Morus bassanus*. However, birds that live on water but apparently do not need to see through the surface (e.g. European Shags *Phalacrocorax aristotelis* and Manx Shearwaters *Puffinus puffinus*) do not have many receptors containing red droplets.

Such interpretations, however, do not seem to square with the presence of the red droplet field that looks downwards in Rock Doves. While wild Rock Doves may fly over the surface of the sea, they do not need to see through it in the way that Gannets do. Others have suggested that the red field of doves enhances the detection of objects against a green background, but this does not sit comfortably with the predominance of cones containing red droplets in birds associated with looking into water.

Clearly, other functional explanations are needed. However, that these patterns are not random does suggest that they are correlated with specific perceptual challenges posed by life in different environments, and with the conduct of different tasks. Perhaps answers will be forthcoming when more is known about the surfaces and objects that particular species are trying to discriminate when carrying out key tasks.

Distribution patterns of ganglion cells

Patterns in the absolute densities of photoreceptors and retinal ganglion cells are captured in isodensity maps (Chapter 3). These cell density patterns are overlaid on the patterns of cone receptor types and do not mirror them. An example of the way that patterns of ganglion cells and of receptor types are unrelated in the same retina can be seen by comparing the distribution of ganglion cells in the Rock Dove (Figure 3.5) against that of receptor types in the same species (Figure 4.5)

It is clear that the areas of high receptor and ganglion cell densities are regions of the highest spatial resolution in an eye. Highest cell densities usually project in the direction of the axis of the eye's optical system. This is the direction of the highest image quality and is often associated with a structure called a fovea. In foveas the outer layers of the retinal cells are displaced to form a small pit or depression in the surface of the retina (Figure 4.6). Displacing the layers of the retina in this way ensures that the highest-quality optical image is received directly by the photoreceptors, thus avoiding any image degradation that might be caused by passing through layers of the retina. Foveas are not unique to birds and have been recorded in all of the main vertebrate taxa.

When the ways that these regions of high cell density project into the world are considered, it is clear that highest spatial resolution is typically associated

Peregrine Falcon

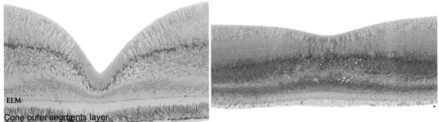

50 microns

Harris's Hawk

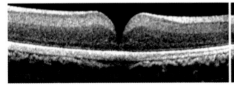
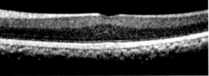

Central deep foveas   Shallow peripheral foveas

FIGURE 4.6 Sections through the retinas of a Peregrine *Falco peregrinus* and a Harris's Hawk *Parabuteo unicinctus*, showing foveas. In both species there are two foveas. One is placed more or less centrally in the retina and looks out laterally in the bird's field of view. The other is placed towards the periphery of the retina and looks more forwards, although it does not look directly forward. The central fovea is deep and has sharply curved sides; this is the location where the highest acuity occurs. The shallow fovea has lower acuity but is still a location where acuity is enhanced compared with the rest of the retina. Note that light making up the image travels from top to bottom in these diagrams and the photoreceptors are in the bottom layers in each illustration. They can be clearly seen in the picture of the Peregrine's deep fovea (top left). (Illustrations courtesy of Mindaugus Mitkus (Peregrine) and Simon Potier (Harris's Hawk), both of the University of Lund Vision Group, Sweden.)

with viewing towards the horizon. Furthermore, these regions usually look out laterally (on both sides of the head) and only slightly forwards. They do not project directly forward in the direction of travel. We are familiar with the fact that we have a single region in which we best see the world. Furthermore, that region lies directly ahead of us, in the direction in which we move. It is perhaps difficult to comprehend that in most birds there are two separate regions of highest resolution projecting laterally on either side of the head. Unlike ourselves, a bird wishing to see something in detail usually takes a sideways look at it.

Distinct patterns of ganglion cell density have been described in a wide range of species, and we have already seen two isodensity maps in Figure 3.5. Alongside the Rock Dove, two further examples are shown in Figure 4.7. Patterns may be roughly circular with a very high concentration at the centre and a gradual decline of cell density away from the centre. In some species more than one region of high concentration can occur. This is seen in Rock Doves, where a region of high

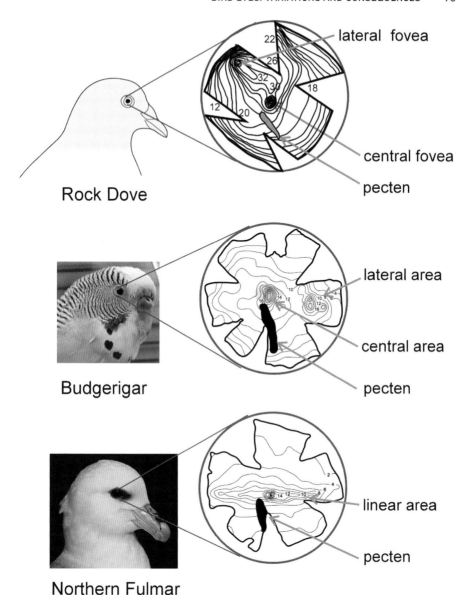

FIGURE 4.7 Patterns of ganglion cell distribution in Rock Dove *Columba livia*, Budgerigar *Melopsittacus undulatus*, and Northern Fulmar *Fulmarus glacialis*, showing some of the diversity that has been recorded. These are isodensity contour maps (see Figure 3.5) and numbers indicate the density of receptors in 1000s per mm². (The diagrams of Fulmar and Budgerigar are from Mindaugus Mitkus of the University of Lund Vision Group, the diagram of the dove is redrawn from work published by Bingelli and Paul. Photo of Budgerigar by Michael Cole, Fulmar by Steve Garvie.)

density projects directly out from the centre of the retina, and hence laterally with respect to the bird's head. It would project more or less straight out of the page. Another centre of cell concentration projects downwards and more forwards, but not actually in the direction of the bill.

In Budgerigars, as in Rock Doves, there are two areas of elevated acuity produced by localised high concentrations of photoreceptors. There is a central region which projects laterally from the bird's head and straight out of the page at the reader, but the second area projects backwards in the bird's field of view. It is presumably associated with enhancing the detection of predators, and/or perhaps with keeping an eye on the activities of other birds in the flock. Budgerigars are highly social birds.

In other birds, high densities of receptors may occur in a linear band (linear area) which stretches across a portion of the retina. In some species this linear area stretches across the full width of the retina. Linear areas have been described in seabirds which forage in the open ocean, including Manx Shearwaters (Figure 3.5), Leach's Storm Petrels *Oceanodroma leucorhoa*, and Northern Fulmars (Figure 4.7). However, extensive linear areas have also been described in terrestrial birds of open habitats, such as Ostriches.

Such linear areas of high cell density are typically positioned so that they project approximately horizontally when the bird's head is held in its usual resting or flight posture. They may be concerned with fixating the horizon, but they may also be for gaining greatest detail in the direction in which other foraging birds might occur close to the sea surface. Detecting other foraging birds may be a vital cue when a shearwater, or other seabird, is searching for places where prey is concentrated in the open ocean. A linear area that projects diagonally across the field of view has also been described in Canada Geese *Branta canadensis*, Snow Geese *Anser caerulescens*, and Greater Flamingos *Phoenicopterus roseus*, but the exact projections of the linear bands of high acuity when the birds adopt different postures are not known in detail, and so their functions are not well understood.

New patterns of ganglion cell and receptor densities, and the positions of foveas, are regularly described. For example, descriptions of ganglion cell distributions have recently become available in various species of waterfowl (ducks, geese), galliforms (quail, pheasants), owls, Cathartidae (New World vultures), parrots, passerines, and penguins. It is clear that these patterns do vary markedly between species, and there is good evidence of some variation between individuals within species, suggesting that these patterns could provide the basis for intense and rapid natural selection tuned to different tasks.

Understanding the function of ganglion cell distribution patterns in the visual ecology of a particular species is often not straightforward. As more patterns are described so it becomes increasingly difficult to find clear patterns that point towards a single function, or a narrow range of functions. But this may be because not enough is known of the behaviour of the species and the visually guided tasks that they carry out.

It does seem, however, that areas of highest resolution have a particular role in the foraging of most birds, especially in the detection of individual items in the lateral field of view. For example, a foraging thrush may cock its head and apparently use lateral vision to seek items on the ground; a falcon or eagle may appear to scan around and use lateral vision when looking for a distant target.

Areas of high-resolution vision may also be used for the detection of birds of the same species. For example, high-resolution lateral vision may be essential for detecting the movements of other birds when flying within a flock, as in Budgerigars. In other species it may function primarily for monitoring the behaviour of other individuals that are foraging at the same time, since their behaviour might give clues to rich foraging patches, or to the presence of a predator, as in seabirds. There does not, however, seem to be any evidence that suggests that the directions of high-resolution vision have a specific role in the control of locomotion. Control of locomotion will be discussed in the next chapter.

The ways in which receptor cell and ganglion cell distributions can vary within a retina, plus the marked differences between species in these patterns, certainly attest to the fact that birds of two different species, placed in exactly the same position at the same time, are very likely to retrieve different information from the world about them.

Human eyes have a distinct region of high acuity and enhanced colour vision, but this region looks directly forwards, quite unlike any bird yet described. Why humans have eyes that look directly forward may be more to do with the use of vision to guide the manipulation of objects held in the hands, rather than looking for distant objects that lie ahead.

These differences between ourselves and birds are quite fundamental and question our ideas of reality. Birds and humans certainly see the world differently, such that even in the same situation they will be extracting different information from their surroundings. Although they are different, each set of information is as real as any other.

Comparing the performance of eyes

Given all that has been stated above about subtle variations between eyes, is it possible to compare them in a straightforward way? Clearly, there is plenty of scope for variation in the structures of eyes of different species. These variations can be intriguing in their own right especially if they can be related to the behaviour and ecology of different species. But is it possible to quantify differences in vision across species in a way that allows rapid comparisons? Can they be compared to provide insights into our everyday observations of what birds of different species do? Ways to quantify sensory capacities using training techniques were described in Chapter 2. What insights do they provide into how vision differs from one species to another?

Acuity

The easiest measure for comparing vision between species it to determine the limit of detail species can detect, that is, their highest spatial resolution. This is measured using high-contrast black and white stimulus patterns. With standard procedures and stimulus patterns, it is possible to be confident that the same ability is being tested in each of the species, and so valid interspecies comparisons of this key aspect of vision are possible. High-contrast stimuli of the kind used in these behavioural tests are in a sense artificial, for they rarely occur in nature. However, acuity measured in this way gives an indication of the very best spatial discrimination of an eye.

Another way of estimating acuity was described in Chapter 3. It combines information on the focal length of the eye's optical system (which gives an indication of the size of the image) with information on the highest density of retinal ganglion cells (as described in the 'Distribution patterns of ganglion cells' section above). Such acuity measures have the advantage of being obtained relatively rapidly compared with the training techniques. However, they do require eyes from recently dead birds and so they are often done using eyes from birds which have died as a result of trauma or disease.

Because the two techniques have been shown to give comparable results in the same species, it is possible to combine acuity measures that use either technique into a single table, as shown in the Appendix at the back of the book. Acuity measures are available for a wide range of species which differ in their evolutionary origins (13 avian orders), in their general ecology, and in their behaviours. For example, the list includes seed- and fruit-eating passerines, scavengers, diurnal raptors, and nocturnal owls.

In the context of all that has been said above about the possible sources of variation in eyes, it is not surprising to find that acuity does show marked differences across species. For quick reference Table 4.1 has been abstracted from the Appendix. It shows that acuity in birds ranges over at least 30-fold. The highest acuity is found in diurnally active raptors (Wedge-tailed Eagle and Indian Vulture) and the lowest in small passerines (House Sparrow and House Finch). It should be noted that the highest acuity so far recorded in any animal is that of Wedge-tailed Eagles and Indian Vultures. It is twice the highest acuity recorded in young humans and about five times that of the standard to which human vision is usually corrected by spectacle lenses, which represents average acuity among older humans. It is also worth noting that falcons (Brown Falcon) have acuity about half that of an eagle, the same as the best young humans.

This wide range in acuity across bird species is remarkable, and it reinforces the point that when viewing the same scene different species will have different information available to them. It may also be surprising to some readers that the acuity of eagles and falcons is not higher, at least compared with humans. Historic estimates of the acuity of eagles based on anecdotal observation have often been used to suggest that the acuity of eagles is more than 10 times that of humans.

TABLE 4.1 Spatial resolution in a sample of bird species and in young humans. Resolution is shown as both cycles per degree and minutes of arc. The two measures are interchangeable; older studies tended to use acuity, while more recent studies prefer to express resolution in cycles per degree. All behavioural measures were made at high daytime light levels and indicate the best performance (finest spatial detail that can be resolved). To compare between species simply divide their acuity values. Thus, the eagle's spatial resolution is about 8 times higher than that of a dove, and 30 times higher than a sparrow's. Expressed another way, an eagle should be able to detect a given object 8 times further away than a dove, and 30 times further away than a sparrow.

Species	Spatial resolution (cycles/degree)	Acuity (minutes of arc)
Wedge-tailed Eagle *Aquila audax*	142	0.2
Indian Vulture *Gyps indicus*	135	0.2
Brown Falcon *Falco berigora*	73	0.4
Rock Dove *Columba livia*	18	1.7
Canada Goose *Branta canadensis*	9.6	3.1
Great Horned Owl *Bubo virginianus*	7.5	4.0
House Sparrow *Passer domesticus*	4.8	6.3
House Finch *Haemorhous mexicanus*	4.7	6.4
Great Cormorant *Phalacrocorax carbo* (underwater)	3.3	9.1
Human *Homo sapiens*	72	0.4

This has gained wide currency, but there really are no data to support it. Although the definitive work on eagle vision was published more than 30 years ago it has yet to gain wide currency.

To give context to the comparison between eagles and humans it is important to consider the function of high resolution in birds, and to consider just what a two- or four-fold difference in acuity means in the natural environment. The important point to consider is what high resolution might be used for. We may tend to think that high acuity is good for seeing small things close by. In our everyday lives this might be the case – after all, without high acuity you would not be able to read this page. For birds with the most acute vision, however, high resolution probably does not function to aid in the detection of small objects close by, but is important for the detection of larger objects at a great distance. This certainly seems to be the case in the larger raptors, including eagles and vultures.

For these birds, food items are rare and scattered widely across the environment, but they are relatively large. It seems likely that it is the need to detect such objects at great distances that has led to the evolution of the highest-acuity vision. The acuity of a Wedge-tailed Eagle crucially means that these eagles can detect a given object at five times the distance of an average human observer, and twice that of a young person with the keenest vision. Thus, an eagle is certainly able to see an object of interest long before a human. Anecdotal observation of this might lead to exaggerated conjecture about the 'super-sense' performance of the eagle. The bird certainly out-competes the young human, but not by as much as sometimes conjectured.

The eagle's acuity is at least eight times better than a dove's. Therefore, a dove would have to be at least eight times closer to the same object for it to be detectable, but it would not in any case be interested in the same targets as the eagle. Clearly doves would gain no advantage in being able to match the performance of an eagle or even of ourselves, since their ecology does not require the detection of objects at large distances. Thus, it would seem that doves or sparrows, which feed on small items detected on the ground at close range, have evolved much lower spatial resolution. Eagle-eyed acuity would have no selective advantage for doves or sparrows. However, the higher resolution of a falcon would mean that the bird could have detected its prey and be well on the way towards it at high speed, long before the prey has detected that the predator is coming.

Although the highest acuity of humans is 0.4 minutes of arc, this is found in younger eyes. It is noteworthy that an acuity of 1 minute of arc is regarded as 'normal' or 'adequate' for humans to complete everyday visually guided tasks, including reading this text. The resolutions of many bird species cluster around this value. In humans, spectacle corrections are usually prescribed to achieve such a level of acuity. Acuities lower than 1 minute of arc are for clinical purposes bracketed into broad categories. Acuity between 1 and 3 minutes of arc is regarded as 'mild vision loss', between 3 and 8 minutes of arc it is termed 'moderate visual impairment', and acuity lower than 10 minutes of arc is labelled 'severe visual impairment'. Thus, some of the birds listed in Table 4.1 and the Appendix (notably the Galliformes, Anseriformes, Columbiformes, and some of the Passeriformes) would be regarded by human standards as having mild vision loss, while the owls have between mild loss and moderate impairment.

Acuity and light level

An important consideration is that the acuity measures of Table 4.1 and the Appendix apply only at the highest natural daytime light levels. They indicate the best possible resolution of each species' eyes but, as noted in Chapter 2, there is an important trade-off in vision between resolution and sensitivity. This trade-off is made clear in the way that in all vertebrate species resolution falls as light levels decrease.

Acuity has been measured across the full range of naturally occurring light levels in only a few animal species, but it does include some birds and, of course, humans. In all species a decrease in acuity with light levels is clearly seen (Box 2.3). In all species, including eagles, falcons, doves, chickens, and owls, acuity drops rapidly as light levels decrease towards twilight. Furthermore, this decrease accelerates as light levels drop within the night-time range. It should be noted that the scales in the graph in Box 2.3 are logarithmic, that is there is a 10-fold difference in magnitude between each value on the vertical axis. This is for convenience, since if linear scales were used it would be difficult to fit the data on the page because the magnitude of acuity changes are so large.

The decrease in acuity within the nocturnal light level range is very steep both in humans and in Great Horned Owls *Bubo virginianus*; it is not so steep in Barn Owls *Tyto alba*, but the decrease is still appreciable. Clearly it would be useful to have acuity–luminance functions of the kinds shown in Box 2.3 for more species. However, producing such data demands many months of investigation using training techniques. Unfortunately, they cannot be determined from eye structure, as is possible for estimating maximum acuity.

It is important to note that over the thousand-fold range that the high light levels of daytime naturally vary, spatial resolution does change considerably, especially as light levels fall towards twilight. This is something that we are all well aware of and is why we reach for the light switch or car headlights as twilight approaches. However, over the million-fold range of light levels that can occur during night-time, the decline in visual resolution is precipitous.

These changes in acuity with light level have the important consequence that bird species whose activity is restricted to higher daylight (the majority of species) will experience relatively stable spatial resolution. That is, acuity will vary little during the birds' active periods, and it will vary little from day to day. The consequence is that for daytime birds there is relative stability in their ability to gather visual information from the environment.

On the other hand, nocturnally active birds, especially the owls of woodlands, such as Great Horned Owls and Tawny Owls, are unable to extract such fine spatial information. Furthermore, the nocturnal birds will also experience large and unpredictable changes in their ability to retrieve spatial information from night to night. This is because factors such as the presence and phase of the moon, or the presence of cloud cover, appreciably alter light levels.

Absolute visual sensitivity

In the comparison between the eyes of doves and shearwaters (in the 'Image production: optics of the eyes of doves and shearwaters' section above) it was stated that the image in shearwaters' eyes was brighter than in doves' eyes. This attests to the fact that there is scope for natural selection to alter the absolute sensitivity of an eye, the minimum amount of light that can be detected.

As with acuity, absolute sensitivity of a species is best measured using a behavioural technique. In effect we want to pose the question, 'What is the minimum brightness of a light source that a bird can detect?' The best kind of light source for this purpose is a trans-illuminated screen which the bird has to compare with an un-illuminated screen. As explained in Chapter 2, this is a difficult task to do. It requires a lot of training, and various tricks are necessary to maintain the motivation of the bird being tested (or even a human, for that matter) so that it keeps responding when asked to function close to the limits of what it can see.

Similar to acuity, the absolute sensitivity of an eye can be influenced by both image production (optics) and image analysis (the retina). Optics will determine the light-gathering capacity of the eye and hence the brightness of the image, while the greater the number of photoreceptors, in this case rod receptors, the higher the probability that light photons in the image are detected and signalled to the brain.

As regards optics, one prediction is that eyes which have evolved to maximise light capture should be absolutely large, since this allows the entrance pupil to be large, which particularly increases the chances that point sources of light will be detected. However, the focal length of the eye should not increase in proportion to the overall eye size. This is because the important determinant of image brightness is the *f-number*.

F-numbers will be a measure familiar to photographers who are concerned about the 'speed' of their lenses. The f-number is the ratio of the focal length of the optical system to the entrance pupil diameter; the lower the f-number the brighter the image. Both large eye size and low f-number are indeed found in the eyes of owls, suggesting that natural selection has operated to maximise sensitivity in the eyes of these birds. As explained above, the eyes of owls are so large that, unlike other bird species, they do not sit within the protection of the skull in a deep orbit, but instead are attached almost to the surface of the skull (Figure 4.2).

Analysis of the optical system of Tawny Owl eyes shows that not only are the eyes absolutely large, but they have a relatively low f-number. The overall size and focal lengths of the eyes of Tawny Owls and humans are very similar, yet they differ in the maximum brightness of the image which they produce. The f-numbers of human and owls' eyes are 2.13 and 1.30 respectively. The result of this is that when viewing the same scene, the image in an owl's eye is approximately 2.7 times brighter than in a human eye (it is the square of the f-numbers which must be compared).

At first sight this difference may appear rather small; however, it is functionally significant. It agrees very well with the differences in the absolute sensitivity of Tawny Owls and humans tested using the same behavioural training technique. Furthermore, the absolute sensitivity of Tawny Owls is 100 times higher than that of Rock Doves.

These comparisons are important, since they show that the difference in absolute sensitivity between humans and owls can be accounted for by the greater light-gathering capacity of the owls' eyes, suggesting that the sensitivity of the retinas is very similar in both species. Indeed, both the owl's retina and the peripheral parts of the human retina have high concentrations of rod receptors. On the other hand, the 100-fold difference in sensitivity between owls and doves is not explained by differences in image brightness (the f-number of dove eyes is in fact very similar to that of human eyes), and it must be attributable to differences in the sensitivities of their retinas. As explained in the 'Variations in the distributions of photoreceptors and ganglion cells' section above, the retinas of doves are dominated by cone receptors, reflecting that their activity is restricted to high daytime light levels.

These comparisons of absolute sensitivity, like those of acuity, show clearly that both the image-producing and the image-analysing systems of bird species differ in remarkable ways. Furthermore, they show that these differences are concerned primarily with the sensory challenges posed by life in different environments.

Shades of grey and colour

Natural stimuli are rarely black and white. Most real-world tasks require the detection of lower-contrast targets, targets that are made up of shades of grey, or targets that differ in colour. An animal's ability to detect these more natural stimuli can be described by measuring *contrast sensitivity functions* and *chromatic acuity*.

Contrast sensitivity functions determine the minimum amount of contrast that can be detected when a bird is presented with patterns of stripes (gratings) in shades of grey rather than black and white. Generating contrast sensitivity functions using behavioural training is a very time-consuming process but it has been achieved in some birds. Like the acuity–luminance functions of Box 2.3, it requires repeated measurement of visual thresholds, and this requires special techniques and patience to maintain the bird's motivation.

From contrast sensitivity functions it is possible to determine the smallest contrast that is visible at a range of stripe widths. Thus, a grating pattern is presented and the contrast between the stripes is changed until it is not possible to detect the pattern. This defines the contrast threshold for that particular stripe width or spatial frequency. The test is repeated for each of a range of stripe widths. Examples of such functions are shown in Figure 4.8. These show that birds have surprisingly low contrast sensitivity compared to mammals, including humans. That is, at a given stripe width contrast has to be higher for birds than for mammals before stripes can be detected. This is indicated in Figure 4.8 by the fact that the contrast sensitivity functions of the birds all sit below and inside that of humans.

Figure 4.8 also shows that contrast sensitivity declines as stripe widths both increase and decrease away from a central range in which contrast sensitivity is highest. This means that there is a mid-range of stripe widths where small amounts of contrast differences are relatively easily detected. However, when stripes are either closer together or wider apart, higher contrast is needed for their detection. What this indicates is that in the natural world both very small and very large objects have to contrast highly with their backgrounds in order for them to be detected.

The overall conclusion is that each species has an optical system that equips it best to detect natural objects within a relatively narrow range of sizes. It is possible that this reflects the kinds of objects that different species seek when foraging, but it is not clear exactly which tasks might drive these differences in contrast sensitivity between species. It is important to note, however, that bird species certainly do differ markedly in their sensitivity to contrast. They also have different contrast sensitivity to that of humans. This low contrast sensitivity of birds must

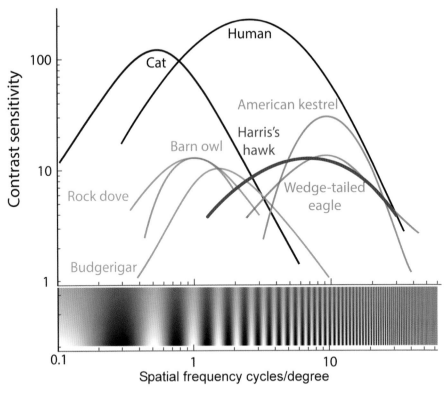

FIGURE 4.8 Contrast sensitivity functions of birds, humans, and cats. To determine these functions the animals are presented with grating patterns of different spatial frequency (stripe width). At each frequency (as indicated across the lower part of the diagram) the contrast of the pattern is varied and the minimum contrast that the bird can detect at that frequency is determined. In the case of Harris's Hawk, a training procedure similar to that shown in Figure 2.7 was used. The threshold contrast at each frequency tested is then turned into a measure of contrast sensitivity to give an average contrast sensitivity function for the species. It is clear that for each species there is a range of spatial frequencies in which sensitivity to contrast is high. However, sensitivity drops off rapidly at both higher and lower frequencies. It is clear from this diagram that bird species differ markedly in their sensitivity to contrast, and that all birds are much less sensitive to contrast than humans and cats. This means that scenes have to have higher degrees of contrast for birds to be able to see detail within them compared with ourselves. (Redrawn from an original figure by Simon Potier of the University of Lund Vision Group.)

play a part in how light conditions determine when a species should begin and end its foraging activity.

Despite the apparent elaboration of bird retinas for the provision of colour vision (Chapter 3) it is perhaps surprising to find that there has been little investigation of how colour influences spatial resolution in birds. Evidence on the spatial resolution of coloured patterns in birds is limited to just two species, Budgerigars *Melopsittacus undulatus* and Harris's Hawks *Parabuteo unicinctus*. Despite the ubiquity of coloured objects and patterns in the natural world it is surprising to find that information is available on these abilities for just two other species in the entire animal kingdom, Honeybees *Apis mellifera* and humans.

The common finding across all of these species, including the birds, is that spatial resolution for coloured patterns is considerably lower than for black and white patterns. This means that, for example, a pattern of green and red stripes would only be detected at higher stripe widths than could be detected if the pattern was black and white. In other words, fine colour patterns cannot be so easily detected as black and white patterns.

It is also clear that there are differences between the two species of birds investigated so far. Harris's Hawks show finer chromatic resolution than Budgerigars. However, with data for only two species it is premature to conjecture on the functions of these differences, but it could point towards colour being a more important cue in the foraging or social behaviour of Harris's Hawks than in Budgerigars. This may seem a bizarre conclusion, given the difference in the colour contrast in the plumages of these two bird species. However, it may only mean that the subtle colour patterns of plumage are detectable at close range rather than at a distance.

As will be discussed in later chapters, these general findings on how spatial resolution declines with ambient light levels, and with contrast, have clear consequences for the sensory ecology of birds. This is especially so when birds are active in situations in which ambient light levels are low and/or variable, or when tasks involve the detection of low-contrast targets. Such conditions are, of course, typical of the real world in which birds are active.

Resolution and ultraviolet

The ability of birds to detect UV light is often remarked upon, and it is frequently presented as something rather special, as one of nature's gee-whiz phenomena. There is, however, an important link between spatial resolution and the ability to detect UV light that presents a different perspective on the ability to detect UV. This is due to the phenomenon known as chromatic aberration.

Chromatic aberration results in a blurred image. It arises because light of different wavelengths is not brought to the exact same points of focus by an optical system. The precision with which light can be brought to a focus by an optical system is inversely related to wavelength. This means that light of shorter wavelengths (towards the blue and violet end of the spectrum) is more spread out, or less well focused, than light from the red end of the spectrum.

Therefore, there is always good reason to filter out UV light if the sharpest image is to be gained. Indeed, this is what occurs in many optical devices, including cameras, and in many eyes, including our own. In many eyes the cornea and/or lens have the ability to filter out UV. Thus it might be expected that eyes which have the highest resolution should remove UV from their image. There is good evidence that this does in fact occur. Diurnal raptors (Accipitridae, Falconidae), which have the highest acuity, are among those birds which do not have UV sensitivity, and it has been shown that their lenses and corneas filter out UV light from the focused image. Birds which have UV sensitivity (e.g. songbirds, ostriches) on the whole have relatively lower acuity and hence the slightly blurred images caused by the presence of UV may not compromise their acuity, as they would not be detectable. It should be noted that the widely reported claim that falcons have UV vision, which they use to guide their foraging, is no longer supported. This will be discussed in more detail in a later chapter.

Chapter 5

Visual fields

The previous chapter showed that differences in the optics of birds' eyes influence two key ways that vision can differ between species: sensitivity and acuity. But optics also influences a third important parameter of vision, the visual field of the eye. This is important because the visual field defines the space from which an eye can gain information at any one instant. Furthermore, all birds have two eyes and their fields combine to give the bird's total field of view. This total field is key in understanding the behaviour and ecology of a bird; it has important consequences for how birds detect predators, food objects, other birds, and obstacles.

Visual fields can differ markedly between species, because eyes can be placed in the skull in many different ways. Although eyes are always placed symmetrically on either side of the head, they can take up many different positions, more forwards, more sideways, tilted upwards, tilted downwards, and their fields can overlap by various amounts. The result is that there are many different bird's-eye views. This chapter is about understanding these different views, their consequences, and the factors that account for them.

From our human way of looking at the world some of the visual fields of birds may be difficult to envisage or comprehend. Particularly problematic to envisage is the world experienced by those birds which have total panoramic vision above and around the head. Such comprehensive visual coverage occurs in only a few species, including some familiar ducks and shorebirds, but just imagining what the world is like from their perspective is quite a challenge. While this might be a world view of only a handful of bird species, the visual world of every bird species is much more extensive than our own. In many birds just a small head movement is all that is necessary to check on what is behind or above them. In some birds, their visual fields allow them to see directly downwards beneath their bill while they are standing upright. However, there are other bird species that cannot even see their own bill. Clearly, there are many different ways in which birds see the world and pick up information on what is around them at any instant.

At first consideration some of the visual fields found in birds may seem bizarre. However, from a broad comparative viewpoint it is perhaps humans who should be considered strange. Our arrangement of two eyes placed in the front of the skull is certainly uncommon, shared primarily with other species of primates. The great majority of animal species have eyes more on the side of the head, which means that in most animals, not just birds, the eyes do not look directly forwards.

Forward-facing eyes perhaps give humans a perspective in which we experience the world as always lying ahead, not above us or behind us, and hardly to the side. This gives us the impression that we are constantly moving forwards into the world. In the majority of animals, and certainly in most birds, the world surrounds them, and they flow through it. As they move through their world, many birds can track an object from in front of them, past them, and watch it retreat behind them – something that is impossible for us, but which we would no doubt find very useful on occasion.

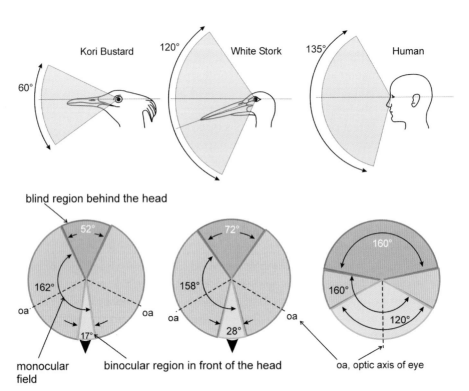

FIGURE 5.1 Depictions of the visual fields in bustards, storks, and humans. The top half shows diagrams indicating the vertical extent of the binocular fields, and the lower half shows sections through the visual fields in a horizontal plane when the heads are in the positions shown. As discussed in the text, there are many differences in these fields of view. However, the most important point to note is that the visual fields of individual eyes in all three species are very similar, approximately 160° in width, but in each species the fields of the two eyes are combined in different ways. The result is that the total fields about the animals' heads differ markedly. For example, there are clear differences in the width and height of the binocular fields, and in the size of the blind areas above and behind the head. The binocular regions are shown in green, the blind region in grey, and the regions served by one eye only in orange. The direction in which the birds' bills project is indicated by a large arrowhead.

Our human arrangement of forward-facing eyes is particularly strange because while each eye has a coverage of 160° in the horizontal plane, their frontal position means that our total horizontal coverage is only 200°, with about 120° seen by both eyes at the same time, i.e. with binocular vision (Figure 5.1). Compare this with the examples of two bird species in which each eye also has a field of view of similar width to ourselves. In Kori Bustards *Ardeotis kori*, the two eyes are combined to give a total horizontal field that is 308° wide. In White Storks *Ciconia ciconia*, total coverage is about 290°. Furthermore, the binocular portions of their fields are markedly different from our own, with maximum widths of only 17° in bustards and 28° in storks. There are also marked differences between these bird species in the vertical extent of the region of binocular coverage and where it is centred with respect to the bill (Figure 5.1). Storks have vertical coverage of 120°, while Kori Bustards have only 60°, and humans 135°. Clearly, eyes with visual fields of the same width can be combined in different ways to produce quite different visual coverage of the world about an animal's head.

These differences are more fully captured by diagrams which show the total visual fields about the head (Figure 5.2). These diagrams are rather complex because they try to capture the full three-dimensional complexity of visual fields. The reader needs to imagine that in each of these diagrams the bird's head is at the centre of a sphere and the features of the visual field are projected out and onto its surface. The different portions of the visual fields are marked in different

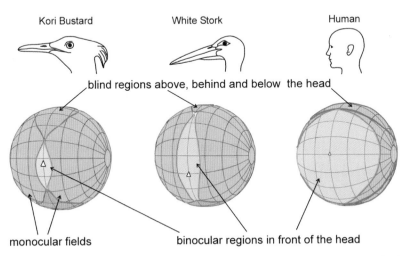

FIGURE 5.2 Depictions of the total visual fields of bustards, storks, and humans in which the key features of the field are shown as projected onto the surface of a sphere surrounding the head. This presentation emphasises how different the visual fields of these three species are. This is despite the visual fields of their individual eyes being similar. It is clear that differences in eye position in the skull, as well as differences in the visual fields of individual eyes, can result in quite different ways in which birds can extract information from the world about their heads.

colours, as in Figure 5.1, and conventional latitude and longitude coordinates are used to mark out the surface of the sphere.

Complex as these diagrams are, they give the best indication of how visual fields differ between species, and they allow easy and swift comparisons to be made. Casual comparison of the full fields of these two birds and humans shows just how different is the space around the head from which information can be obtained at any one moment in each of these species. This is all the more remarkable because the eyes in all three species have visual fields of almost the same dimensions. The fields of the two eyes are combined in radically different ways in each species, to give radically different world views. An obvious and important difference between ourselves and these birds is that while we have extensive binocular vision, we also have very extensive blind areas above and behind the head. The two birds see much more of the world than we do.

It is important to note that in both the bustards and storks, and for that matter all birds, forward vision is achieved by using the periphery of the visual field of each eye (see Figure 3.3). Each eye is looking outwards to the side, which means that their optic axes diverge, and only the periphery of each eye looks forward (Figure 5.1). This is in marked contrast to humans, in which it is central vision that looks in the forward direction. The optic axes of our eyes project forwards, parallel to each other. In birds the central axis of each eye always looks laterally to some extent. Even in raptors, including owls and hawks, which are often incorrectly referred to as having forward-facing eyes, the visual axes diverge by at least 50°.

Comparison of visual fields and their functional interpretations has provided a rich source of ideas in visual ecology. Not only do visual fields show some very marked differences between species, they also show some subtle differences between closely related species. Interpretation of these differences shows that visual fields, like other aspects of vision, are subject to constant evolution through natural selection, and that they are driven by and tuned to key tasks in the daily lives of birds.

General characteristics of the visual fields of birds

In birds, placement of the eyes on the sides of the skull means that each eye looks outwards at a different scene, and there is little overlap in the visual fields of the two eyes. This means that, as shown in Figures 5.1 and 5.2, the binocular fields of birds are relatively small and are typically between 20° and 30° wide, but they are vertically long. They extend to 180° in some birds, including some species of herons, ducks and sandpipers. However, such vertically extensive binocular fields are typically very narrow, only 5–10° in some ducks (Anatidae, e.g. Mallard *Anas platyrhynchos*) and some sandpipers (Scolopacidae, e.g. Eurasian Woodcock *Scolopax rusticola*). See the example, shown in Figure 3.4, of a duck which has narrow binocular vision both in front of and behind its head. In fact, the binocular field in these birds extends through 180° from in front of the head, over the top of the head, to directly behind.

The broadest binocular fields found in birds are typically between 40° and 50° wide; they are found in some songbirds, owls, and some hawks. The widest binocular fields reported in birds to date are in fact found in crows (Corvidae) with the widest in New Caledonian Crows *Corvus moneduloides*. At 60° wide they are still only half the width of the human binocular field, and the eyes in these crows are certainly not forward-looking. The broader binocular field width in this particular crow species may have a specific function in these birds' use of tools for obtaining food items; this is discussed in more detail below.

In many birds the eyes not only look sideways from the skull, but they are also positioned more towards the top of the skull with the axes of the eyes projecting slightly upwards. In other species the eyes point slightly downwards, with the result that when they are standing upright, and the head is held horizontal and slightly forwards of the body, they are able to examine objects at their feet, this is found in herons and egrets (Ardeidae) (Figure 5.3). Extensive vision above the head is not found in all bird species. Some, for example ostriches, raptors (eagles, falcons, vultures, and owls), albatrosses (Diomedeidae), and hornbills (Bucorvidae and Bucerotidae), have a broad blind area above the head.

The direction in which the bill projects within the visual field also varies markedly across species. In some species the bill falls squarely in the centre of the binocular field, as seen in herons (Figure 5.3). In other species the bill projection falls below the centre of the visual field, as seen in storks and bustards (Figure 5.1), but all

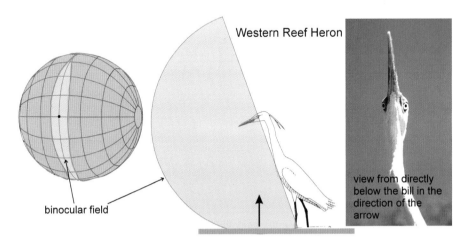

FIGURE 5.3 The visual field of a Western Reef Heron *Egretta gularis*. Herons have very extensive visual field with only a small blind area behind the head. Most of the world is seen by either the left or the right eye alone. The binocular region is narrow but extends vertically through 180°. The result is that when standing upright with the bill horizontal a heron can see what is at its feet. The photograph on the right captures this well, showing the heron apparently looking at the camera which is directly below the bird.

of these species can probably see their own bill tips. However, in some birds the projection of the bill falls outside or at the very edge of the visual field, and the bill tip cannot be seen. This is the case in some ducks, including Mallards, which, as described above, have binocular fields that extend above and behind their heads. In short, they can see all around them, but they cannot see their own bill.

What shapes the visual fields of birds?

It seems that both the general and the detailed features of bird visual fields have been shaped primarily by the key perceptual challenge posed by foraging. For most birds this challenge is the accurate positioning of the bill (but in some species the feet) when taking food or prey items. However, accurate positioning is not just to do with placing the bill in the right place. It also involves getting the bill to the target at the right time. This allows the bird to seize something in its bill or feet with high accuracy.

We may be impressed by the ability of a falcon to stoop on prey and snatch it in mid-air, or an eagle's ability to capture a fast-moving hare on the ground. However, these are no more impressive than a pigeon pecking at a small food item (Figure 5.4). Both require being spot-on to the target's position and knowing exactly when the target will be reached. Pecking may seem to be a simple everyday action which does not need explanation compared to the more dramatic pounce of an eagle. However, getting either of these actions right requires the same precise information from the visual system: information about where an object is and how long it will take to reach it.

FIGURE 5.4 A Wood Pigeon *Columba palumbus* pecks at grain, a Golden Eagle *Aquila chrysaetos* pounces on prey. Which species faces the more exacting task? The pigeon must bring its bill accurately to the grain and the bill must be opened at the correct time to seize it. The eagle must align its talons towards its prey and the talons must be spread and then closed with precision timing. In both instances the task is basically the same. The birds must achieve correct direction of travel and accurate prediction of time of arrival. (Wood Pigeon photo by mike193823319483; Golden Eagle edited from an original video frame provided by Simon Baxter.)

Pecking, lunging, or capturing prey with the feet has to be got right day in and day out throughout a bird's life. And for many birds this has to be achieved while also keeping an eye out for potential predators. In essence, a stooping falcon or a pouncing eagle has it easy compared with a pecking dove. The falcon has few predators, and all it needs to do is fix and follow its prey and ensure that it can overhaul it. A pecking dove, on the other hand, needs to fix the target and gets its bill to it while at the same time gaining information about any nearby potential threats.

Accurate bill (or foot) placement in space and time, while simultaneously detecting predators, is a task that most birds have to achieve almost constantly, certainly repeatedly, every day. There are, however, exceptions – and they are instructive. Some birds do not need to place their bills accurately, and the form of their visual fields helps us understand the selective factors which have shaped visual fields in general.

As well as accurate bill placement and predator detection, there is a further factor that has shaped visual fields, but perhaps only in certain species. This is the need to avoid imaging the sun in the eye. For humans, keeping the sun out of our eyes seems to be essential, not only to reduce the possibility of long-term damage to the retina and optical system but also to protect images from glare effects. We go to great lengths to avoid imaging the sun; we naturally use our eyebrows to shade the eyes, but headgear that shields the eyes has existed since antiquity. The problem is that if a bird has vision over a wide field of view above and behind its head, it is difficult, if not impossible, to avoid imaging the sun. There is clear evidence that avoiding seeing the sun has been crucial in the evolution of the visual fields in birds with large eyes. On the other hand, birds which have smaller eyes seem able to tolerate seeing the sun.

What might seem surprising in the above paragraphs is that no mention is made of using vision to control locomotion (flight or running). Certainly, there is ample evidence that birds do use vision for these purposes. However, controlling locomotion does not seem to be a primary factor in shaping birds' vision.

Seventy-five years ago, birds were famously described by André Rochon-Duvigneaud as 'wings guided by eyes'. He coined this phrase in the introduction to his survey of the eyes and vision of vertebrates, which at the time of its publication in 1943 was a state-of-the-art summary of what was known. The phrase seemed to capture the essence of birds: flight and vision were combined in just a few words, and it emphasised that vision was the key to the spectacle of bird flight. That phrase has often been repeated. However, it now seems that a bird is more appropriately described as 'a bill guided by an eye'. Perhaps this phrase conjures up a less exciting idea of what a bird is, and what a bird does. To emphasise control of the apparently mundane task of pecking over the control of aerial acrobatics may seem strange. However, the perceptual requirements for the control of locomotion may not be particularly exacting and they seem to be met within the requirements of vision that is shaped by foraging and predator detection (Figure 5.5).

FIGURE 5.5 Flight, foraging, or predator detection? Which is the key task that has shaped the vision of birds? Is the driver of vision the control of flight, controlling the position of the bill, or detecting predators? All three tasks are of importance throughout the daily lives of nearly every bird, but which is more important, which tasks have driven the evolution of vision in birds? (Photo of New Zealand Kākā *Nestor meridionalis* in flight by Maree Mcleod [CC BY-SA 4.0], Kākā foraging by Kate Macbeth [CC BY-SA 4.0], Black-tailed Godwit *Limosa limosa* in flight by Hans Hillewaert [CC BY-SA 3.0], Godwit foraging by Charles J. Sharp [CC BY-SA 4.0], White-chinned Petrel *Procellaria aequinoctialis* in flight by J. J. Harrison [CC BY-SA 3.0], Petrel foraging by Mjobling [CC BY 3.0].)

The key functions of bird visual fields

A number of strands of evidence support the idea that controlling bill position, including the accurate timing of its arrival at a target, is the most demanding task that vision is used for by birds. The second most demanding task is the detection of predators. However, these tasks make competing demands, and the configurations of visual fields are primarily the result of this competition.

Control of bill position

Regardless of whether food items are taken by pecking or by lunging with the bill, or by seizing an item in the feet, the essential requirement is to arrive accurately at the right place at the right time. It is perhaps surprising to find that controlling the direction of travel towards a target and determining time to contact with that target do not require information on either absolute distance or relative depth. This may seem counterintuitive. Surely, we need to at least know how far away something is in order to be able to determine how long it will take to get to it? This is certainly true when we are planning a journey, or when the distance to an object is relatively large, say 100 m. However, for targets at short range or when we are travelling at high speed relative to them, the computation of distance by the visual system is in fact too slow for determining time of arrival. Computing distance and depth takes time, and although the brain can work very fast, this is not fast enough in a dynamic situation. For example, managing to catch a ball, or working out just how to bring a vehicle to a halt at a stop line, requires that accurate information on position and time to contact must be gained very rapidly. The same considerations apply to a bird when timing the arrival of its bill or feet so that it can seize an object. In these situations, speed of approach can be very rapid. Hence both direction of travel and time of arrival have to be literally spot-on if foraging is to be successful.

We are often fascinated by demonstrations of relative depth using the process of stereopsis (stereoscopic depth perception). In this, the brain compares the slightly different views of the same near object which arise in each of our eyes. There are various devices (generally referred to a stereoscopes) which use this to render an illusion of a scene in 3D when viewing a pair of images taken from slightly different viewpoints. The trick of these devices is ensuring that one eye sees only one image and the other eye sees only a slightly different one of the same scene. The brain tries to make sense of these different images and interprets it as objects lying at different depths in front of us.

Fascinating as these demonstrations are, stereopsis actually works properly only with close objects and small relative depths. Stereopsis does not give us accurate depth perception for objects even a few metres away. For that, we use different cues concerning, for example, the relative sizes of known objects, how they overlap one another, and how their relative positions change as we, or they, move. These processes work well enough to give a sense of the space around us, but they work best for relatively static scenes.

If we are in a dynamic situation, in which either we or the objects in the scene are moving rapidly, we rely upon a very different set of information. This is gained from what is known as the optic flow-field (Figure 5.6). Information extracted from flow-fields is thought to underpin important aspects of dynamic vision in most species, including invertebrates such as bees and flies, as well as vertebrates, including birds and humans. All natural vision is in fact dynamic, caused by the motion of the observer or motion of objects within the world about the observer; only under controlled laboratory conditions can vision be thought of as static.

There is in fact very scant evidence that birds have stereoscopic vision. It may have been demonstrated in Barn Owls, but even in them it would work only at very close range. This is because the eyes are very close together in the skull and so the two views of the same object lying ahead are almost identical. Very high acuity would be needed to detect the differences which are the essential basis for stereopsis. Humans have the advantage of a big head with eyes that are relatively

FIGURE 5.6 Optic flow depicted using arrows surrounding the point to which an eye is travelling. The images of objects within the field of view flow at different velocities across the retina depending upon how far away they are from the observer, and where they lie within the field of view. The point to which the eye is moving is stationary and the flow pattern expands symmetrically about that point. Images of objects near the observer flow faster than images of objects further away. The relative speed of flow is indicated here by arrow size: the larger the arrow, the faster the image is flowing across the retina at that point. (Photo courtesy of Max Planck Institute of Neurobiology / Schorner.)

far apart, giving rise to more readily detected differences in each eye's view of a close object.

Optic flow-fields describe the ways in which the image of the world moves across the retina as the head moves through space. It has been regarded as the foundation of visual perception. It was first described, and its implications explored, nearly 70 years ago by the American psychologist J. J. Gibson, who went on to propose a radically different framework, which he called ecological optics, to describe and investigate visual perception. The fundamental role of flow-fields in the control of various aspects of locomotion, including flight behaviour and the timing of approach to a target, have now been well established in both vertebrates and invertebrates.

A key property of optic flow-fields is that when moving directly towards a target the flow-field expands symmetrically about the very point to which an observer is moving, and thus the focus of the flow pattern specifies exactly where an animal is heading; that is, it specifies the direction of travel (Figure 5.6). Furthermore, the rate at which the flow-field, or the image of the world, expands across the retina specifies directly how long it will take to contact the object. There is no need to have information about how absolutely far objects are away. If they are far, then the flow-field expands slowly, if they are near it expands rapidly, but both rates of expansion predict how long it will take to reach the target.

We use this kind of information from flow-fields to control most, some would argue all, important aspects of our everyday actions. It has been shown most conspicuously to be used when doing exacting tasks such as car driving, running and jumping, or catching a ball – tasks for which instant access to information on direction and time to reach a target are essential. These are tasks which are too immediate for the brain to determine the distance and the speed of travel, and then to compute the time to arrive at a target. Only the expanding flow-field can provide the information sufficiently fast.

In birds, a target is usually first detected visually in the lateral field of view of a single eye. This will be done employing a region with the highest-quality optics and the highest retinal resolution, often involving a fovea (the the 'Distribution patterns of ganglion cells' section in Chapter 4 above). For example, we may often see a thrush foraging on the ground tilting its head to look for a worm using its lateral vision, or we may see a falcon scanning around for a target, apparently using lateral vision (Figure 5.7). However, after detection, visual control has to be passed to the frontal part of the visual field. It is at that time that our casual observations suggest that the bird starts to look at the target – but of course it has been looking at it long before.

What the bird is doing when moving from viewing the object laterally to frontally is not gaining a more detailed view of the object, in fact the detail it sees might decrease. However, by this move the bird is bringing the image of the object into the region of binocular vision. This is the part of the field of view in which the optic flow-field expands symmetrically about the target that the bird is aiming for.

FIGURE 5.7 An American Robin *Turdus migratorius* foraging on a lawn, a Peregrine Falcon *Falco peregrinus* scanning the sky for possible prey. Both are using lateral vision to guide the task of prey detection, employing the region of highest acuity within their fields of view. Only when prey is detected and the bird needs to close in on it is visual control switched to forward vision, which is used to guide the bill or the feet towards the prey. (Photo of American Robin by Geoffrey E. Hill.)

A bird that pecks for its food opens it bill to seize the item just a split second before grasping it. This split-second timing is also seen when an eagle pounces on its prey (Figure 5.8). An eagle approaches its prey at high speed with the feet trailing behind and below its body. The head is aligned so that the image of the prey item falls within its binocular field. It is only at the last moment before impact that the feet are brought forward and the targeted prey, feet, and binocular field are brought into alignment (Figures 5.4 and 5.8). This alignment of the head/bill/feet towards the prey also ensures the accurate timing of their arrival and means that talon spreading is precisely coordinated with arrival at the prey.

Predator detection

Having complete visual coverage of the world all around the head would seem to be the ultimate adaptation for meeting the demands of predator detection. This is, however, found in a relatively small number of bird species. If it can occur in some species, why does it not occur in all? Many birds have extensive visual fields, but most have a blind area behind the head, and this must leave them more vulnerable to predator attack compared with species that have comprehensive vision.

The presence of blind areas in most bird species, and their absence in only certain species, provides convincing evidence on the demands that ultimately shape vision in birds. It seems clear that accurate control of bill position, and the detection of predators, are the twin key tasks of vision. However, they make different demands for information, and these demands are in competition.

Total panoramic vision has evolved independently in two quite different bird orders: ducks (Anseriformes) and shorebirds (Charadriiformes). Even in these taxa panoramic vision is not common. However, those species which have panoramic vision, such as Mallards *Anas platyrhynchos*, Shovelers *Spatula clypeata*, Pintails *A.*

FIGURE 5.8 A Golden Eagle pounces on an artificial moving prey object. These three photographs are taken from frames of a single video sequence. The binocular portion of the visual field has been superimposed in light blue. The sequence shows that the eagle has the prey more or less centrally in its binocular field all of the time. As it approached the feet are held back beneath the body. At the very final moment the feet are swung forward into the binocular field and directly in line with the moving target. As the target is grabbed the feet are placed centrally within the binocular field. Although this sequence shows the view from the side, a frontal view also reveals that when the talons are brought up and spread wide just before impact, they too are placed within the binocular field, as shown in Figure 5.4. (Sequence courtesy of Simon Baxter, still images edited by the author.)

FIGURE 5.9 An adult Woodcock has retrieved an earthworm from beneath the ground, having detected it using non-visual cues from its bill (see Chapter 7). A Woodcock's chick hatches at an advanced stage of development from a simple nest scrape on the ground, and it does not require feeding by the adult. This tactile foraging technique, plus the absence of the need to use vision for nest building and chick provisioning, means that the evolution of visual fields in woodcocks has been driven primarily by the requirement to detect predators rather than for guidance of the bill. Thus, Woodcocks are among the small number of bird species which have comprehensive vision around and above their heads. (Photo of adult Woodcock by Ronald Slabke [CC BY-SA 3.0], Woodcock chick by Janet Pesaturo, One Acre Farm.)

acuta, and Woodcocks *Scolopax rusticola*, share a common feature. Their foraging does not require visual control of bill position, relying primarily on tactile, not visual, cues (Chapter 7). Furthermore, all of these birds have precocial young (Figure 5.9).

Precocial young are chicks which hatch at an advanced stage of development; within a day or two of hatching they forage and feed themselves and are dependent on their parents only for shelter and protection. This means that adults do not have to find food and place it accurately into an open gape, as is common is most birds. Furthermore, their nests are also simple structures, not requiring fine manipulation and interweaving of materials. Basically, those shorebirds and ducks which have panoramic vision do not need to see and time their bill placement with high accuracy. The ducks dabble and the woodcocks probe blindly, relying upon tactile and taste cues to find food items. For this reason, the bills of these ducks are placed at the very periphery of their visual fields and in Woodcocks the bill tip falls outside their visual field.

It seems that these species are freed from the requirement to gain visual information for the accurate control of bill position, and natural selection has favoured the evolution of comprehensive visual coverage to aid predator detection. This suggests that if information is required to control bill position then it is paramount. The eyes have to be positioned more forwards, so the visual field encompasses the direction of the bill, with the result that in most birds there is a blind area behind the head.

It is important to note that in the duck and woodcock species with panoramic vision the width of binocular overlap is minimal. In order to be able to see behind and above the head the fields of each eye barely overlap. In these species the width of the visual field of each eye is about 180°. This is probably close to what is optically possible in a camera eye without causing distortions in the periphery of the field. The eyes are also oriented slightly upwards, and this gives the birds a binocular overlap of about 5–10° which starts at the bill and extends through 180° to directly behind the head. This means that binocular overlap in the forward direction, in the direction in which they move, is less than 10°. However, these birds are capable of fast flight in complex habitats such as woodlands and often fly when light levels are low. This shows that a frontal binocular field just a few degrees wide is sufficient for the control of flight.

Differences in visual fields between closely related species

There is evidence from both ducks and shorebirds that the features of visual fields can evolve relatively rapidly. Significant differences in visual fields are found between closely related species that must have diverged from a common ancestor relatively recently. Furthermore, these differences in visual fields have been correlated with significant differences in behaviour. Well-studied examples involve differences between pairs of species among ducks, ibises, and shorebirds.

Northern Shovelers and Eurasian Wigeons (Figure 5.10) were until a few years ago placed in the same genus, indicating that they probably diverged from a common ancestor relatively recently. Therefore, from an evolutionary perspective any differences between them, including their vision, must be of recent origin. The foraging of Wigeons is mainly by selective grazing guided by visual cues. They actively seek out growing tips and certain plants within a grass sward and target individual items; this requires pecking that is targeted with some precision. Shovelers, however, do not even need to see what they are foraging on. This is

FIGURE 5.10 Northern Shovelers *Spatula clypeata* feed primarily by filter-feeding. This means that during feeding their bill position does not require precise visual guidance. Like other duck species, their nests are not complex structures and their young are precocial self-feeders. This combination, as in Woodcocks (Figure 5.9), has resulted in Shovelers having comprehensive vision around and above their heads, driven primarily by the need to detect predators. Their bill falls at the very edge of their visual field. Eurasian Wigeons *Mareca penelope*, on the other hand, while having similar nesting and chick characteristics as Shovelers, are visually guided in their foraging. This requirement for precise visual guidance of the bill has resulted in a visual field in which the bill is more central placed in the binocular region, and the binocular region is slightly wider than in Shovelers. As a consequence, Wigeons have a blind region behind their heads. These are relatively subtle differences between the visual fields of two closely related species. However, they lead to Wigeons spending less time feeding and more time in vigilance behaviour, looking around for predators, than Shovelers. (Photos by Peter Blanchard.)

because they use a filter-feeding technique and rely primarily upon tactile and taste cues to detect and select items.

Shovelers have comprehensive visual coverage of the celestial hemisphere about their heads but can only just see their bill tips (Figure 5.10). Wigeons, on the other hand, have a narrow blind sector to the rear of their heads but a wider binocular field which embraces the projection of the bill tip. This suggests that they position their bill tip using visual cues from within the binocular field. Thus, the differences in the position and size of their binocular fields reflects the different foraging behaviours of these two ducks. Moreover, the difference in the extent to which their vision covers the space around the head is reflected in differences in their vigilance behaviour. Wigeons frequently break off foraging and raise their heads, checking for predators. Shovelers do this much less frequently, presumably because their comprehensive coverage of the space around their heads means that they can remain vigilant for predators even when foraging with their heads down near the water surface.

Thus, two closely related duck species, which can often be observed exploiting different resources in the same locality, differ in their visual-field configurations, foraging technique, and vigilance behaviour, and these are all interrelated. Similar differences in visual fields and behaviour have also been found between more distantly related species of ducks (Blue Ducks *Hymenolaimus malacorhynchos*, Pink-eared Ducks *Malacorhynchus membranaceus*), which also differ in their use of visual and tactile cues when foraging. Pink-eared Ducks feed in a similar way to Shovelers, while Blue Ducks feed by taking invertebrates from rock surfaces, beneath stones, or by snatching them from the water column. This must require the same kind of accurate control of bill position and timing that Wigeons have to use.

Among sandpiper species (family Scolopacidae), significant differences in visual fields have also been recorded that are correlated with differences in the extent to which species use tactile or visual cues in prey detection and capture (Figure 5.11). Red Knots rely mainly on tactile cues when feeding on the muds of estuaries during the winter months (Chapter 7). However, in summer when on their breeding grounds in the Arctic tundra, they employ vision to control bill position for taking surface-living and aerially active insects.

Capturing these moving insects demands accurate visual information to position and time bill-opening. Not surprisingly their visual fields show the same kinds of features found in other precision peckers and lungers. They have a narrow binocular field surrounding the bill and a small blind sector behind the head. This is, of course, quite different to the closely related woodcocks, which have comprehensive coverage of the world with no blind area behind the head (Figure 5.9). Woodcocks rely exclusively upon tactile cues to find prey throughout their annual cycle and do not need to use vision to control their bill position. What this comparison also reinforces is the conclusion that it is the more demanding visual task, the task of getting the bill accurately to a target, that is the main driver of visual field configuration.

FIGURE 5.11 Red Knots *Calidris canutus* have two distinct foraging behaviours which come into play at relatively distinct parts of their annual cycle. On their breeding grounds in the tundra they feed primarily on surface and flying invertebrates. Outside of the breeding season they feed primarily by probing in the soft muds of estuaries and seashores for buried invertebrates. This foraging is guided by tactile cues from their bill tips (see Chapter 7). Taking invertebrates when in their breeding-ground habitats is more visually demanding than probing, requiring precision location and timing of the bill. It is this much more demanding visual requirement that has resulted in these 'tactile' feeders having visual fields with characteristics usually found in visually guided foragers. (Photo by Brad Winn, Manomet.org.)

A final example of how small differences in foraging ecology may be reflected in the visual fields of closely related species is found among ibises (Threskiornithidae). Neither Northern Bald Ibises *Geronticus eremita* nor Puna Ibises *Plegadis ridgwayi* have comprehensive vision. They do, however, differ in the extent of their binocular fields, and the size of the blind sectors behind their heads, in a significant way.

Northern Bald Ibises forage in terrestrial habitats and peck food items from dry surfaces and occasionally probe with their bills into fissures. Compared with Puna Ibises, they have a broader binocular field that surrounds the bill, and a broader blind area behind the head. The Puna Ibises probe for items in soft substrates, often through water in marsh habitats, and so they rarely see a prey item and must always rely on tactile cues from their bill tips. The Bald Ibises, on the other hand, may use tactile cues when probing under rocks but they often take individual items, such as grasshoppers, from the surface. Catching these is a demanding visual task requiring precise location and precise timing in the opening of their bill tips.

Bill control versus predator detection

The preceding examples lend support to the key idea that the forms of visual fields in different bird species are driven primarily by the requirement to provide information to guide the bill, which is traded off against the requirement for predator detection. Total panoramic vision about the head is not unique to birds; it is found in some insects and in some mammals, most notably among the Lagomorpha, rabbits and hares.

Among birds, not requiring visual cues to guide foraging is not, however, sufficient to result in the evolution of comprehensive vision about the head. It is also necessary that the bill does not require fine visual control for any task, not just for foraging. The result is that comprehensive vision is found only among birds which also do not need to position their bills accurately for two other key tasks: nest construction and the provisioning of young. Both ducks and shorebirds use simple nests which do not require elaborate construction, and their young are precocial. Most other birds must use their bills for foraging, for nest building, and for the provisioning of young, all tasks which require accurate position and timing of the bill.

One further, and particularly telling, example that makes this clear is provided by filter-feeding flamingos (Phoenicopteridae). Filter-feeding in flamingos is achieved by highly specialised structures in their bills. An enlarged tongue trapped in the mandible pumps water into the mouth cavity from around the bill tip. The water is expelled out at the sides of the bill opening, where it is strained through filter structures. These filters are able to trap minute resources: invertebrates, algae, and diatoms. On the face of it, such feeding would not seem to require fine control of bill position. Furthermore, with the head inverted in its characteristic foraging position, the bill floats more or less at the water surface, and so visual guidance is not even required to position the bill before filtering can start.

The surprise is that flamingos do not have comprehensive vision (Figure 5.12). Their binocular field encompasses the bill tip, which they can see, and they have a blind area behind their head. The reason for this visual field configuration is that while flamingos do not need vision for foraging, they do need to place their bills accurately for two other key tasks. First, they build nests which are placed atop a

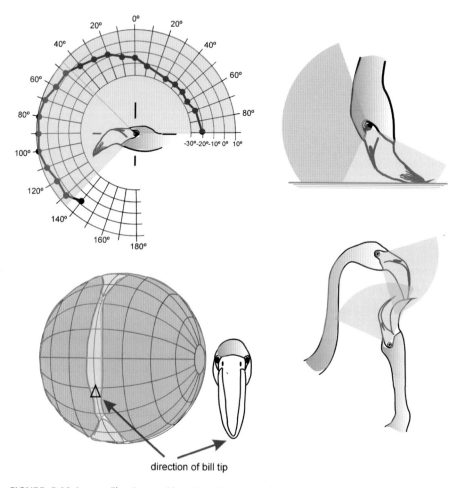

direction of bill tip

FIGURE 5.12 Lesser Flamingos *Phoeniconaias minor* have a visual field whose essential features are much like those found in birds which need to position their bill with precision. They have a vertically long and narrow binocular field (shaded green) with the bill tip projecting within it (bottom left), and an extensive blind area (shaded grey) above and behind the head (top left and top right). At first this is a surprising finding, because flamingos feed primarily by filter-feeding, and when they have their heads inverted at the water surface, they are blind in the direction in which they are heading as they walk forward (top right). However, feeding their young on 'crop milk' requires their bill tips to be placed very accurately: they need to see just where they are placing their bills before transferring milk to their chicks and juveniles (bottom right).

carefully fashioned mound of mud, and second, they feed their young in a highly specialised way that requires their bill to be placed with high accuracy.

Flamingo chicks are fed on 'crop milk' (a secretion from the oesophagus) which is dripped by the adults into their open mouths. The young are fed by both parents in this way for many weeks before the chicks' bills develop so that they are able to start filter-feeding themselves. Thus, despite their filter-feeding, which presumably does not require any visual information for its control, flamingos require vision that allows accurate bill placement so that young can be provisioned. Thus, in flamingos, feeding the young is probably the key driver of their visual-field configuration.

It is also worth noting that the blind area behind the heads of flamingos means that when they are feeding with their head upside down between their legs, they cannot see directly ahead of them. To gain coverage of the world ahead they swing their heads from side-to-side as they walk forwards. This head swinging may also serve to stir up food at shallow depths or within mud.

Visual fields, eye size, and imaging the sun

The emphasis in the above discussions has been on how the control of bill position, and the detection of predators, place competing demands on the essential information that eyes must provide birds. While it is the trade-off between these demands that shapes the basic feature of visual fields, there is third important factor which shapes the upper part of the visual field, the part that looks skywards.

There is good evidence that eye size and the width of a blind region above the head are related. The explanation for this relationship lies in the sun, specifically the problems that arise when the sun is imaged upon the retina. The sun is by far the brightest object that we naturally encounter. In humans, the problems of imaging the sun are generally recognised to be three-fold: there can be temporary damage to the retina from even brief direct exposure to the sun, the lens and cornea can be damaged by exposure over a long period, and there can be direct disruptive effects to vision caused by glare and afterimages which leave temporary 'holes' in the field of view. All of these can be debilitating and can threaten our wellbeing and safety from moment to moment, and over the long term. Not surprisingly humans go to quite some lengths not to image the sun on their retinas. However, it is surprising to find that birds with smaller eyes must be imaging the sun on their retinas for much of their waking time.

It could be that smaller-eyed birds can position themselves so that the sun is always imaged at the periphery of the visual field. This means that the effective aperture of the eye for the sun's image is a slit rather than a circular pupil. This would result in low image brightness. So the problem may not be as great as it first appears, although lifetime exposure of the cornea and lens is potentially a major problem, but perhaps most birds do not live long enough for the damaging effects to accumulate. However, some longer-lived birds do develop cataracts or clouding of their lenses, and this may be due to long exposure to the sun.

Large-eyed bird species are the ones which have a broad blind area (up to 80° wide) above the head. These include eagles, vultures, ostriches, hornbills, and albatrosses. Furthermore, these large-eyed birds also have various structures outside of their eyes (optical adnexa) which can function as 'sunshade' devices. These include brow ridges and thick and elongated eyelashes (Figure 5.13). Such structures are absent in other birds. In fact, it has been suggested that birds can be divided into two groups: 'large-eyed sun avoiders', which have sunshades and a blind area above the head, and 'smaller-eyed sun viewers', which have no sunshade devices and no blind area above the head.

The reasons for this dichotomy may lie in the reasons for having a large eye in the first place. As argued in Chapter 4, in birds that are active in daytime, large eyes probably evolved primarily to give high spatial resolution. However, large eyes also have large entrance apertures and so will produce a relatively brighter

FIGURE 5.13 Sunshades in birds. Griffon Vultures *Gyps fulvus* have prominent ridges of skin which can place the eyes in shadow on the brightest days and give the birds their famous 'hooded' look (top left). Common Ostriches *Struthio camelus* also have extensive brows that shade the eyes plus long eyelash-type feathers (top right). Southern Ground Hornbills *Bucorvus leadbeateri* do not have particularly prominent brow ridges but they do have numerous long eyelash-type feathers which cast shadows across the eyes (lower left and right). (Photo of Vulture by A. Román Muñoz Gallego, University of Malaga; photos of Ground Hornbills by Heinrich Nel, Mabula Ground Hornbill Project.)

retinal image. In humans the image of the sun on the retina is in fact sufficiently bright that it can act as a secondary light source within the eye, scattering light and degrading contrast and hence resolution across the whole of the retinal image. We may also notice this if we look at the beam of a torch or the flash from a camera. It is not just the bright image that knocks out our vision; we cannot see anything much across the rest of the field of view, because light is bouncing around inside our eyes. This is referred to as veiling glare.

Clearly, if selective pressure has been to evolve a large eye to maximise resolution, it would be maladaptive to compromise that resolution by degrading the retinal image with light scattered from an image of the sun. Smaller eyes, on the other hand, can only ever have lower acuity, and so veiling glare produced by light scattered from a retinal image of the sun may do relatively little to degrade the image and decrease spatial resolution across the visual field.

While there are gains in having a blind area that stops the eye imaging the sun in eyes that have high resolution, there is one serious cost which has emerged only recently. This is a cost that has arisen due to human interference in the natural world. Surprisingly, the extensive blind area above the head can render these birds particularly vulnerable to collisions, especially with human artefacts such as power lines, wind turbines, and even hang-gliders. The sensory ecology of such collisions will be discussed in the final chapter.

What is binocular vision used for?

In humans, stereoscopic vision is often regarded as synonymous with binocular vision. The perception of relative depth using stereopsis for the control of fine manipulations of objects with our hands is usually regarded as the prime function of binocular vision in humans. It has often been assumed that this same function applies to all instances of binocular vision. However, it seems unlikely that this is the case among birds. But if birds do not have stereopsis, what is the function of binocular vision?

In birds it seems that binocularity is, in fact, not a consequence of the requirement for binocularity per se. That is, it is not to do with having two eyes looking at the same object at the same time. In birds it seems that the requirement is for having a portion of the visual field of each eye that looks in the direction of travel. This is because it is only if an eye looks forward, in the direction of travel, that it receives an optic flow-field which expands symmetrically about the point to which the bird is heading. As described above ('Control of bill position' section, Figure 5.6), this symmetrically expanding pattern is very important, since it specifies the direction in which a bird (or its bill) is travelling, and how long it will take to get there.

This requirement for a symmetrically expanding optic flow-field means that the visual field of each eye must have a 'contralateral projection'. That is, each eye must look across the vertical plane that divides the head in half and passes between the eyes and cuts through the bill tip (see Figure 3.3). Inevitably, the projection of the field of each eye across the central plane of the head means that there must be an overlap between the fields of the two eyes, hence there must be a binocular portion of the total field of view. Binocularity in itself may not have a function. It may be more appropriate to concentrate on what each eye can do independently as a result of having a contralateral projection, rather than on what the two eyes might be able to do together. Indeed, there is evidence that a dove can peck at a target as accurately with one eye as it can with two. All it needs is to be able to perceive the expanding flow-field in one eye, and it can direct and time its bill-opening with high accuracy towards a target.

The shapes and sizes of binocular fields: peckers and lungers

In birds which use vision to control bill position when taking food items, the visual fields show a narrow and vertically elongated frontal binocular field in which the bill is placed either centrally or slightly below the centre (Figures 5.1 and 5.2). Such an arrangement is found in a remarkably wide range of bird species that differ both in their ecology and in their evolutionary origins. Binocular fields reach a maximum width in the horizontal plane of between 20° and 30° in most non-passerine species. In passerine species binocular field widths of nearly 60° are found.

Such visual field arrangements are found in birds that feed in many different ways – for example, in species which peck at small mainly immobile items (e.g. Ostriches, Eurasian Stone-curlews *Burhinus oedicnemus*, Rock Doves, Eurasian Wigeons, Southern Ground Hornbills); those which take prey by lunging, or by pursuing evasive prey that are taken directly in the bill (e.g. herons, penguins, cormorants); those which take evasive prey or carrion in the feet (e.g. eagles, vultures); and those which snatch prey from a surface while swimming or flying (e.g. petrels, albatrosses).

This similarity in binocular field configuration across such diverse groups of birds, and across such diverse feeding techniques, suggests a degree of convergence upon a binocular field width that is optimal for a particular purpose. That is, a 20–30° binocular field is as broad as it needs to be in order to fulfil a particular function that is common to all of these species.

In all of the birds mentioned above, although the bill is placed approximately at or just below the centre of the binocular field, there are some subtle variations that reveal fine-tuning of the binocular field to particular visual tasks and ecological conditions.

In the majority of birds studied so far, the bill does not actually intrude into the visual field. The result is that ostriches and herons, for example, cannot actually see their own bill tip, nor can they see what is held in the bill, much in the same way that humans can only just see their nose. On the other hand, in birds that employ precision-grasping the bill does intrude into the visual field, and this means that these birds can also see what lies between their mandibles when an object is grasped. Thus hornbills, which pick up small items with forceps-like action in the tips of their large decurved bills, can visually inspect objects that lie between their bill tips before tossing them to the back of the mouth. This ability to see between the opened mandibles is also found in Common Starlings *Sturnus vulgaris*, which use a specialised technique of 'open-billed probing' to uncover prey in the surface layers of a substrate. Seeing between the bill allows the Starling to examine what is has unearthed before pecking at it.

The shapes and sizes of binocular fields: the special case of tool users

New Caledonian Crows are among the small group of birds which forage using tools. These crows use sticks to extract grubs from cavities in wood which they may detect visually by looking into the grubs' bore-holes. The binocular field of these crows is about 60° wide, the widest yet described in birds. It has been argued that these broad fields are necessary to allow the crows to see with one eye along the line of their stick tool. These tools are held in the tips of the bill and usually propped against the cheek below the eye. The result is that the tool projects sideways rather than directly forward as an extension of the bill (Figure 5.14). The extensive contralateral projection of the field of each eye allows these birds to see along the shaft of the tool and hence to be able to control the position of its tip. Once the tool has entered a hole it cannot be seen with both eyes; it can only be seen by looking along the shaft with one eye. There is no point in having a tool which cannot be accurately controlled. Binocular vision does not help these crows see down a hole, but the contralateral projection of the visual field allows the birds to look down the shaft of the tool and control its placement in the hole.

The shapes and sizes of binocular fields: vertical extent

The vertical extent of frontal binocular fields differs between species (Figures 5.1 and 5.2). Thus, while diverse species have a common maximum binocular field width of between 20° and 30° the vertical extent of these binocular regions may vary between 60° (e.g. bustards), 80° (e.g. ostriches, stone-curlews, hornbills, vultures), 120° (e.g. storks), and 180° (herons) (Figure 5.3). This has consequences for the extent to which the birds can see above and below themselves. Thus, a heron standing with its bill horizontal and the head slightly forward of the body can see what is at its feet (Figure 5.3). Seeing what lies perpendicularly beneath the bill clearly has the advantage that a foraging heron can remain motionless, monitoring what is going on below it, while it waits for a prey item, such as a frog or fish to come within striking range. Such prey have evolved rapid escape responses, and a heron may get only a one-strike chance to catch each item. Therefore, monitoring what is going on below, without having to move the head or body, and waiting for prey to come within striking range, is clearly a significant advantage for a heron.

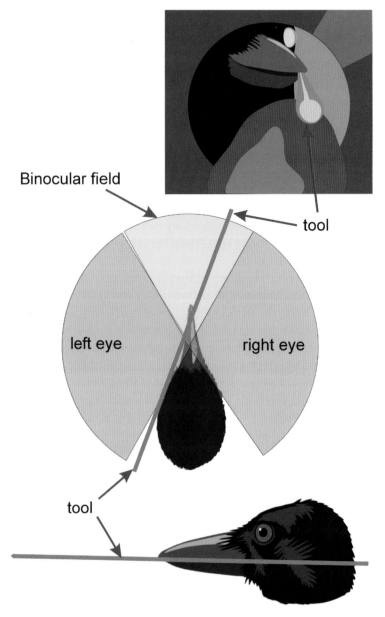

FIGURE 5.14 Tool use and binocular vision in New Caledonian Crows *Corvus moneduloides*. These birds have the widest binocular field yet described in birds. They are famed and well-studied for their ability to fashion and use tools to extract grubs from holes in trees. The holes are narrow, and the birds can usually see down them with one eye only (top). They hold their tool (which is usually straight) firmly by clamping it in the bill tip and propping it against the skull below the eye (bottom). This means that the tools projects laterally, not directly forward (middle). The bird's wide binocular field means that a crow can look along the shaft of the tool and position it carefully when extracting a grub from the bottom of a hole (middle and top). (Based on original drawings by Jolyon Troscianko, University of Exeter.)

The shapes and sizes of binocular fields: nocturnal birds and predators

As pointed out above, the widest binocular fields in birds are not found in owls (Strigidae, Tytonidae) or in diurnal raptors (Accipitridae). There is still a generally held idea that wide binocularity is in some way concerned with nocturnality and predation, but this does not seem to be supported by evidence. It is assumed that binocularity must give some kind of advantage when living life at low light levels. It is not clear at what date the idea that a link between nocturnality and broad binocularity arose. It seems to be based upon casual observations of owls and the assumption that the eyes of owls are frontally placed. The eyes of owls are in fact laterally placed in the skull, although they are more frontal than in other birds.

The maximum binocular field width of Tawny Owls is, in fact, similar to that reported in a number of bird species including the majority of passerines investigated to date. It has been argued that the width of owls' binocular fields may in fact be the product of an interaction between enlarged eyes (associated with maximised light gathering, Chapter 4), and the elaborate outer ear structures, which are unique to owls (Chapter 6). These large outer ear structures may simply prevent more lateral placement of the eyes. Put simply, more extensive visual coverage laterally would not be possible in owls because the outer ear structures would get in the way (Figure 5.15).

It is clear that the visual fields of bird species which forage at nocturnal light levels, or are nocturnally active during key parts of their annual cycle (e.g. Black-Crowned Night Herons *Nycticorax nycticorax*, Paraques *Nyctidromus albicollis*, Oilbirds *Steatornis caripensis*, Manx Shearwaters *Puffinus puffinus*, Woodcocks, Golden Plovers *Pluvialis apricaria*, Red Knot *Calidris canutus*, Stone-curlews, King Penguins *Aptenodytes patagonicus*, kiwi, Black Skimmers *Rynchops niger*) show no evidence that nocturnality is associated with wide binocularity. In all of these species binocular field widths fall within the usual range found in non-passerine species.

Oilbirds are instructive since they are among the most nocturnal of birds (Chapter 9) and their eyes appear to show extreme adaptations of both optics and retina towards increased sensitivity. However, their binocular field width is similar to those of other non-passerine species which are typically active only during daylight.

Summary: the key drivers of vision in birds

Birds clearly use visual information for the control of a very wide range of behaviours. However, there seem to be just two key tasks that have driven the evolution of vision, with all other aspects of visual information obtained within constraints set by these key tasks. The primary driver lies in the perceptual challenges posed by foraging, specifically the control of bill (or feet) position and timing. The second driver is the detection of predators, which requires vision over as wide a sector of

Tyto alba (Scopoli)

FIGURE 5.15 The interaction of eyes and ears in owls. The top illustration shows a Barn Owl *Tyto alba* from which the feathers of the facial disc have been removed to reveal flaps of skin that surround the eyes. These are part of elaborate structures of the outer ears – which are unique to owls and function in the accurate location of sounds (see Chapter 6). The bottom illustration shows a Tawny Owl *Strix aluco*, and sketched over it is a schema which shows the extent of the skull (black) and the approximate positions and extent of the outer ear structures (yellow). It is clear that, as shown in Figure 4.2, the eyes of owls bulge out of the skull and much of what appears to be the head is in fact feathers. It is also clear that in owls their visual fields cannot project far behind the bird's head because the outer ear structures are in the way. (Barn owl illustration by Rolf Åke Norberg.)

space as possible around the head. Getting both of these tasks right more or less continuously throughout the daily life of a bird is crucial for its survival.

These twin drivers, however, make competing demands, and they are best considered respectively primary and secondary. This is because only under the specific circumstance of bill position not having to be controlled by vision does comprehensive visual coverage, the ultimate adaptation for predator detection, evolve. In addition to these two main drivers, it seems that a third driver also operates, at least in eyes beyond a certain size: the need to avoid imaging the sun upon the retina.

There is a further important difference between the two key tasks that drive vision. The control of bill position requires information extracted from the world that lies in front of and relatively close to the bird. The detection of predators, on the other hand, requires information that lies laterally, or even to the rear of the bird's head, and is concerned with information from locations that are remote from the bird. Thus, the regions within the visual field where there is high spatial resolution project laterally on both sides of the bird, not directly forwards. Forward vision is not the region of highest spatial resolution.

There is evidence that predatory birds such as Peregrine Falcons detect their prey at a distance using lateral vision. That is, a target is first detected using a region of high acuity which projects laterally and slightly forwards. When approaching prey, Peregrines frequently do so along a curved path which keeps the prey in the region of highest acuity of a single eye. The birds pass control to the frontal binocular region just prior to prey capture. Such use of lateral vision for detecting food items, with control passing to forward vision for final prey capture in the bill, has been reported in other species. For example, in terns foraging over mudflats for crabs, thrushes searching on the ground for earthworms (Figure 5.7), eagles taking prey in their talons (Figure 5.8), and even in domestic chickens when detecting grains amongst grit.

Conclusion: vision and beyond

Chapters 3, 4, and 5 have been long and complex, and an assiduous reader may feel a bit battered by information and ideas. This reflects both the complexity of vision and the many investigations of vision in birds. Bird vision has been a hot topic for at least a century. I have endeavoured to outline both the basic simplicity of bird eyes, and the wide range of diversity that has evolved in its two basic mechanisms: the imaging (optical) system and the image analysis systems (retina).

Built around common designs, these two systems show great diversity between and within species. In bird retinas, photoreceptors of only a few types are found, but these occur in a very wide range of patterns of abundance in the retinas of different species. It has also been shown that small changes in the physical sizes, positions, and curvatures of the lens and corneas have provided different eyes with images that have markedly different properties.

The descriptions presented here of what is known about the eyes and the vision of birds are not comprehensive. More is known, and a lot more will be known when researchers look into the eyes of more species. However, the examples used show that eyes can vary significantly one from another and, most importantly, that it is possible to interpret much of the diversity by reference to the provision of information for the conduct of specific tasks in different environments.

Specific examples of how this information can be used in understanding the behaviour of birds in different environments are considered in later chapters. Before that is possible, however, it is also necessary to gain understanding of the other senses which provide birds with information about their environments. Vision alone is not sufficient to guide the behaviour of any bird.

Chapter 6

Beyond vision: hearing and smell

Although the previous chapter was all about vision, two other senses, hearing and touch, muscled their way into the narrative. To understand why birds differ in their vision it was shown that information provided by other senses must often be taken into account. For example, we saw that the evolution of panoramic vision in ducks is made possible by their touch sensitivity, while the converse of panoramic vision, the extensive blind areas above and behind the head in owls, is understood by reference to their hearing. These are explicit examples of how senses interact and trade off between one another.

While vision may be key to understanding many actions of birds, it is always important to know what information birds receive beyond vision – what they hear, taste, smell, or feel. All of these senses play key roles in the daily survival of every species of bird. It is important not to think of senses beyond vision as secondary sources of information. Senses beyond vision provide vital information for the control of both the everyday actions of all birds, and the specialised actions of particular species.

The chapters that follow present many examples of how non-visual senses play crucial roles in the lives of birds. However, what is known about these senses is not as great as that available for vision. Hence it is sometimes necessary to rely on broad generalisations across species in order to understand the information that underpins many actions. In later chapters we shall discuss the sensory information that birds depend on to carry out particular tasks or cope with specific environmental challenges. These include foraging below the surface of water, living at low light levels, and detecting prey hidden under a cover of leaf litter. In all of these tasks it will be shown that non-visual information is often essential, and sometimes dominant.

Before such topics can be discussed, it is necessary to gain general insights into what kinds of information the non-visual senses can provide. In this chapter the capacities of hearing and olfaction (the sense of smell) are described. Like vision, these two senses are telereceptive. That is, they are 'far-receiving', able to provide information about objects that are remote from an animal. Sometimes they can detect objects many kilometres away, but usually they gather information about the world no more than a few tens of metres away. Restricted as their range may be, they nevertheless provide information about the wider environment, and this initial information is often checked out or located in detail by vision. A sound or a smell may alert a bird, but checking the exact source may be done by vision.

In Chapter 7 the two other senses that are 'beyond vision', touch and taste, are discussed. These are the key 'near-receiving' senses. They provide information about objects that vision can rarely access or check out. The stimuli that the senses of touch and taste respond to have to be literally in touch with the bird's body. Typically, they provide information used to guide foraging, but they may also play a crucial role in such key behaviours as parental care and intimate adult interactions. Taste also provides information that is crucial to the wellbeing of a bird, especially the identification and the qualities of food items. This is usually with respect to items taken into the mouth, but taste also plays a role after food has been ingested and moved into the alimentary tract. In essence the near-receiving senses provide information about a bird's intimate environment, about objects and substances that are in contact with or very close to the bird's body, and even inside the body.

The final sense to be discussed will be magnetoreception. In the past it has been portrayed as a 'magical' or 'sixth' sense, and it is certainly unlike other senses because it is neither near- nor far-receiving. It is concerned neither with remote objects nor with objects touching the body surface. Magnetoreception stretches our imaginations because it is inaccessible to us, but it is not unique to birds, being found in many other animal taxa. It provides information about ambient conditions and is used by birds to determine their location without reference to objects. It seems to provide birds with a kind of map, their location within it, and a compass to navigate by.

Hearing: information from sounds

The sounds of birds are an everyday pleasure of the natural world. For ornithologists, bird sounds can be an essential tool for their work in the field. This is because bird sounds can contain information about which species are present, how many there are, their sex, and their motivational state. This information is available to the birds themselves, and so hearing plays a key role in regulating the social behaviour of many species. In some species, information from sounds is essential to their foraging, and in a few bird species hearing is used to guide their very movements, through the use of echolocation.

The informational world of sounds is rich and varied. It provides streams of information that are accessible through no other sensory channel. We can intercept some of that information using our own hearing abilities and so, to some extent, we are able to enter into the sound information world of birds. We may even be able to extract richer information from the sound worlds of birds than the birds themselves are able to. This is due to our cognitive (learning) abilities, which give us the ability to learn to distinguish between the sounds produced by a large number of species. The birds themselves are probably extracting sound information that relates only to their own, or close competitor, species.

There seems to be little doubt that in some respects the hearing of birds is inferior to our own and to that of other mammal species. It will be shown below that all birds are able to hear a similar suite of sounds, and that these fall within the range of frequencies and intensities that we can hear; in fact, we can also hear sounds that birds cannot. A further important difference in the hearing of humans and the majority of bird species is that birds are not able to locate the source of a sound with high accuracy. Although locating the sources of sounds may be vital to survival, or for the control of social interactions, many birds cannot do this with high accuracy.

Despite these limitations, sounds provide a unique suite of information that is not available through other senses. Key to this is one crucial property of sounds: they can provide information about sources that are hidden from view. As a result of this, sounds produced by the movements of other animals have become particularly important in the foraging of some bird species.

As a source of information that cannot be seen, sound has also led to the evolution of species-specific sounds that are deliberately produced to mediate social interactions (Figure 6.1). These sounds include, of course, the familiar songs and calls, which can be highly complex, varying rapidly in frequency and in time. They are produced by the specialised vocal apparatus of birds, the syrinx. This is located in the body cavity at the base of the trachea. An elaborated system of membranes under direct muscular control causes air expelled from the respiratory system to vibrate, and this vibrating air is broadcast through the mouth, with the frequencies of sounds often enriched by resonances produced in the mouth and trachea.

There is a range of other sounds produced by birds for communication that are less complex than songs and calls. However, they are no less important in regulating the social behaviour in certain species. Many of these sounds, because of their low frequencies and volume, often carry much further than songs or calls. They are produced by more direct and simpler mechanical means (Figure 6.1) – for example by woodpeckers (Picidae) drumming on wood with their bills, snipes *Gallinago* spp. using specialised tail feathers which vibrate to produce sounds during the dives of aerial displays, pigeons *Columba* spp. clapping their wings together during exaggerated undulations of display flights, and storks *Ciconia* spp. making sounds by exaggerated bill-clapping. In some species, the produced sounds may be amplified by elaboration of anatomical structures such as the long and coiled tracheas found in some species of cranes and swans, or the inflated air sacs used to produce low-frequency booming sounds by some grouse species and herons.

FIGURE 6.1 Birds produce sounds for the purposes of communication using a variety of mechanisms. Songbirds (top left, Eurasian Reed Warbler *Acrocephalus scirpaceus*) produce sounds using the vocal apparatus, while woodpeckers (top right, Great Spotted Woodpecker *Dendrocopos major*) tap out stereotyped drumming patterns on wood using their bills. Other species use their tails or wings to produce sounds during display flights. Common Snipe *Gallinago gallinago* (middle left) extend specialised tail feathers to produce a 'drumming' or 'bleating' sound particularly in steep dives, while Common Wood Pigeons *Columba palumbus* (middle right) bang their wings together during looping flight patterns to make loud and stereotyped short sequences of clapping sounds. White Storks *Ciconia ciconia* (bottom left) clap their bills together during display behaviours which usually take place at a nest site, and many grouse species, exemplified here by Greater Prairie Chickens *Tympanuchus cupido* (bottom right), produce low-frequency sounds employing inflated air sacs as resonators. (Photos of Reed Warbler, White Stork, and Common Snipe by Richard Chandler, Wood Pigeon by mike193823319483, Prairie Chicken by GregTheBusker [CC BY 2.0].)

The sound stimulus

Sounds are transmitted through air more rapidly than odours, but more slowly than light. However, at close range sounds can provide, for all intents and purposes, instantaneous information about nearby objects and other animals.

All physical movements of objects cause the medium that surrounds them (air or water) to oscillate. These oscillations are momentary compressions of air or liquid molecules. The compressions are not static, but propagate away from the object through the surrounding medium in the form of compression waves. Thus, information about the movement of something can be detected remotely from it by detecting these compression waves; this is the very basis of using sound to extract information from the world. A sound tells us that something has moved. It is these compression waves that ears detect, and it is the task of the brain to determine what might be their cause, and where it is with respect to the listener.

Properties of the oscillations caused by the movements of an object are correlated with properties of the object. For example, whether the object has hard or soft surfaces, and how fast and how far it moves, all influence the nature of the oscillations that are emitted. Thus, sounds present a rich source of information about moving objects and events in the environment. These events may vary from some relatively slow but very large-amplitude oscillations of air produced by weather systems, to the very small-scale oscillations produced by the vocal apparatus, or the rustle of vegetation caused by an animal moving across a forest floor.

In absolute terms, the amplitudes of these oscillations of air molecules are very small and of low energy. To detect such minute oscillations, ears have evolved to become super-sensitive movement detectors hidden within the protection of the skull. Movements of air molecules that animals, including all birds and ourselves, routinely detect are smaller than the wavelength of light.

The threshold of hearing in young humans is approximated by the faint sound of leaf-litter rustle heard from a couple of metres and is described as having a sound pressure level of between 0 and 10 decibels (dB). Close to this threshold, the displacements of air molecules at the ear are about the diameter of a hydrogen atom ($\approx 10^{-10}$ m), which is about 1000 times smaller than the wavelength of visible light. Even sounds which we describe as very loud, for example a car horn sounding a couple of metres away (a sound pressure level of about 100 dB), are produced by air molecules whose movements are approximately 10,000 times larger than threshold movements, but they are still only about 1 μm (10^{-6} m) in amplitude. Thus, bird ears are gaining information about the world from their ability to detect extremely small movements of air molecules. Human hearing spans air oscillations which have an amplitude range of approximately 1 million-fold, but at the highest sound levels (140 dB) damage to the ears is highly likely to occur.

Ears can detect these vibrations not only over a range of amplitudes but also over a range of frequencies of oscillation. However, no species has hearing that covers the whole frequency range of sounds produced on Earth, and compared to mammals, hearing in birds is restricted to a relatively narrow range of frequencies. In some mammals, such as dolphins and microbats, hearing extends to frequencies as high as 100 kHz. Among the baleen whales and elephants hearing extends to very low frequencies of about 1 Hz. Bird hearing does not detect sounds at these very high or very low frequencies, and so may be viewed as impoverished compared with that of mammals in general.

Comparing hearing sensitivities

The most convenient way of displaying and comparing the hearing of different animals is to show audiograms (Chapter 2, Box 2.2). These display the minimum sound levels necessary for detecting sounds (the threshold of hearing) at a range of frequencies. Audiograms can be constructed using information gained by a number of different techniques, but the most valuable audiograms are those derived from behavioural studies of the kinds outlined in Chapter 2.

By using training techniques, a bird can indicate that it is able to detect a sound of a specific frequency. The amplitude of the sound that it can just detect defines the hearing threshold. Repeating this across a range of frequencies produces a snapshot of the limits of the individual's hearing, and combining data for a number of individuals produces the audiogram for the species.

As with determining the thresholds for vision, these studies are laborious, and it is often difficult to maintain the motivation of birds when judgements become difficult as sounds approach threshold levels. It is not surprising therefore that only a few detailed behaviourally determined audiograms are available for birds. Furthermore, audiograms show considerable individual variation within a species, but by taking mean values it is possible to compare the average hearing of different species. Audiograms can also be constructed using more rapid physiological techniques, for example recording from the auditory nerve, and on the

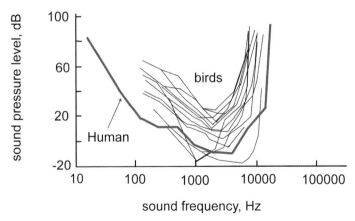

FIGURE 6.2 Audiograms of birds and humans. Audiograms for a number of different bird species are shown together with an average (or standard) audiogram for young humans. Since sensitivity to sounds generally decreases with age, this human audiogram is perhaps indicative of the best performance of human hearing. Many readers will have hearing that is less sensitive and spread over a narrower frequency range. The typical U-shape of vertebrate audiograms (see Box 2.2) is evident in all bird species. This shape indicates rapid loss in hearing sensitivity to sounds of both higher and lower frequencies, but there is a relatively broad frequency range where hearing is most sensitive. The hearing of birds sits within the overall envelope of hearing of young humans, indicating that whatever sound a bird can hear, a human will also be able to detect it. There are no secret frequency ranges used by birds. The sound frequencies of songs and other sounds used for communication (Figure 6.1) sit well within the hearing of young humans.

whole audiograms determined in this way support the data from the behavioural audiograms.

Among birds, average audiograms show considerable similarity across species, indicating that the hearing sensitivities of all birds share similar characteristics (Figure 6.2). All audiograms are U-shaped; there are rapid declines in hearing sensitivity for both high- and low-frequency sounds, and a broad range of frequencies where sensitivity is similar. The general pattern among the audiograms of birds indicates that highest sensitivity to sounds is in the frequency range 1–4 kHz. Towards the high frequencies there is a rapid decline in sensitivity with a cut-off at about 8 kHz, while at low frequencies the cut-off is at about 300 Hz. The absolute threshold (the lowest-intensity sound that can be heard) is at a sound pressure level of about 20 dB in the 1–4 kHz frequency range. This general U-shaped pattern is found in the audiograms of most animal species. It is found in humans, but our hearing extends to both higher and lower frequencies than in birds. However, both humans and birds in general show a similar range of frequencies where they are most sensitive, and absolute thresholds are also similar.

There is evidence that some birds can detect infrasounds (Figure 6.3). These are sounds below the range audible to humans, less than 20 Hz. However, this has not been demonstrated in a detailed audiogram. Sensitivity to such low frequencies

FIGURE 6.3 Three bird species for which evidence has indicated that they are able to detect infrasounds: the familiar domesticated pigeon (Rock Dove *Columba livia*), another domesticated species (Helmeted Guineafowl *Numida meleagris*), and a Golden-winged Warbler *Vermivora chrysoptera*. These three species are from widely separated avian families, which suggests that detecting low-frequency sounds could be widespread among birds. However, it is not clear that this ability to detect low-frequency sounds is a function of the birds' hearing. (Photo of Rock Dove by Muhammad Mahdi Karim, www.micro2macro. net [GNU free documentation licence], Guineafowl by New Jersey Birds [CC BY-SA 2.0], Golden-winged Warbler by Luke Seitz.)

was first reported in Rock Doves, in which it was shown that they could detect sounds with a frequency as low as 1 Hz. The function of this low-frequency sound detection has been linked to navigational mechanisms in Rock Doves, but it may also serve to warn of advancing intense weather systems.

This use of hearing as a weather warning mechanism was suggested to have occurred in Golden-winged Warblers, in which satellite tracking showed that some birds left their breeding grounds ahead of a tornado event and returned when the event had passed. The explanation advanced was that low-pressure weather systems are known to emit low-frequency sounds at high intensity. These transmit over distances of hundreds of kilometres and have been used in human weather forecasting. Infrasound detection has also been reported in Chickens *Gallus gallus domesticus* and in Helmeted Guineafowl, but the function is unknown since these birds do not travel large distances or even fly readily, so it is unlikely to be used for avoiding weather systems.

Interesting as these reports are, it is not clear that detection of such infrasounds does in fact constitute instances of hearing. It could be that such low-frequency sounds, at the high intensities at which they have an effect, are detected by somatic (touch) receptors that provide information about internal conditions within the limbs and the cardiovascular and respiratory systems (see Chapter 7). Thus, when birds sense these low-frequency sounds, they may actually 'feel' them, rather than hear them. There is evidence that this kind of infrasound detection may even occur in humans, with reports of people 'feeling' low-frequency sounds from within their bodies, especially the stomach and gut, sometimes resulting in nausea.

The audiogram of young humans is the typical U-shape with frequency sensitivities cutting off at 20 Hz and 20 kHz, compared with the 300 Hz to 8 kHz range of most birds (Figure 6.2). Thus, young humans can certainly hear all the sounds that a bird can hear, plus sounds with frequencies both above and below the birds' range. It should be noted, however, that human hearing declines with age and a loss of both sensitivity and a narrowing of the frequency range is inevitable, starting from about the age of 18 years. However, even in middle age most humans will still have hearing that embraces the frequency range of most birds. Thus, it can be concluded that the hearing of birds sits within the hearing of the majority of humans.

Locating sounds

Detecting a sound can provide a bird with valuable information about what is nearby. However, the value of that information is greatly augmented if where the sound is coming from, the location of the sound source, can also be determined. It is all very well to know that something is out there making a sound, but knowing its location will often be crucial. The problem is that sound location is not straight-forward. It has two components, direction and distance, and these are determined independently. An ear of itself cannot determine either direction or distance. This is quite unlike vision, where the position of an image on the retina immediately specifies the direction in which an object lies.

The direction and distance of sounds are determined by different mechanisms which have different degrees of accuracy (Figure 6.4). The determination of sound direction is a purely sensory ability that relies on information extracted simultaneously from both ears. Determining sound distance (also referred to as sound ranging) can be achieved with one ear and is primarily a cognitive ability that depends on familiarity with the characteristics of known sounds in different environmental conditions.

These descriptions of what is involved in locating a sound apply to ourselves as well as to birds. There is no simple way for us to directly determine the distance of a sound source; experience and learning are required. On the whole birds are much less accurate than us in their abilities to determine both direction and distance. This may seem surprising given the importance of sounds, especially songs and calls, in regulating the social behaviour of many species.

Determining the direction of a sound source

A single ear can signal only the presence of a sound. Determination of a sound source's direction requires its detection by both ears. It is small differences between the signals received at each ear when detecting the same sound source that are used by the brain to gain an idea of its direction.

Sound direction mechanisms exploit two major sources of information: differences in the intensity, and differences in the time of arrival, of the same sound at each ear. These differences arise because the ears' entrances are on opposite sides of the head, facing away from each other, separated by the width of the skull. This means that the sound is slightly attenuated when it arrives at the ear furthest

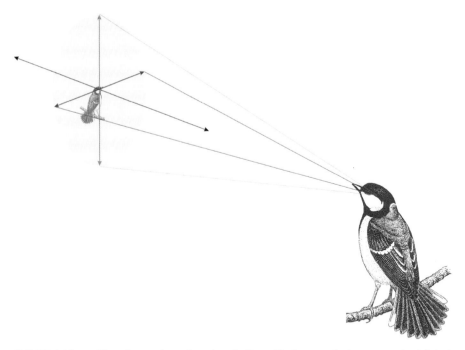

FIGURE 6.4 Sound location and ranging. A male Great Tit *Parus major* hears a territorial rival singing nearby, but it is obscured by vegetation. How accurately can the listener locate the singer? There is a high degree of uncertainty, which arises from two quite different cues. The direction of the singing bird can be anywhere within a large ellipse of uncertainty (yellow shading) with the extent of uncertainty in the vertical plane (blue) greater than in the horizontal plane (red). The distance (range) of the singing rival is determined from other information, and the bird can appear to be anywhere along the black line, nearer or further away than its actual position. Note that the scale of this diagram is greatly compressed; the degrees of uncertainly in location and range are much greater in the field.

from the source, and it will also arrive slightly later. These differences are used to determine sound direction in all terrestrial vertebrates, including birds and humans.

Humans have big solid heads in which the ear entrances are widely separated, and this means that time and attenuation differences are relatively large, giving us the ability to make fine and accurate sound direction judgements. Bird heads are quite different. In all species they are small and not very dense (bird skulls do not have the thick bone of mammal skulls), so sound differences between the two ears of birds are very small, both in intensity and in time. Differences in intensity depend on the frequency of the sound, and in Common Starlings *Sturnus vulgaris* between-ear intensity differences vary from 2 to 8 dB over a frequency range of 1 to 8 kHz. Differences in the time of arrival of sounds at the two ears are about 100 microseconds. Theoretical analysis of these differences leads to the conclusion that most birds will be unable to determine the directions of sounds with high accuracy, and this has indeed been demonstrated in behavioural tests.

Sound direction accuracy: songbirds and Budgerigars

In humans, best sound direction accuracy is between 1° and 2°. In Common Starlings it is between 19° and 27°, and this is not the poorest recorded in birds. In Great Tits, depending on the frequency of the sound, direction accuracy in the horizontal plane varies between 20° and 26°, in Zebra Finches *Taeniopygia guttata* accuracy has been recorded as lying between 71° and 180°, in Atlantic Canaries *Serinus canaria* it is between 49° and 71°, and in Budgerigars *Melopsittacus undulatus*, 25–69°.

These directional location performances are not very impressive. In fact, they might be considered surprisingly poor given the general reliance of these birds on the use of songs and calls in controlling their social interactions. If these angular uncertainties in location are translated to physical distances, their inaccuracy is perhaps even more surprising. For example, a 20° localisation accuracy means that a sound source 20 m away could be located anywhere within a horizontal arc 7 m long, while a sound 50 m distant could be located anywhere in an arc 17 m long.

With so few species studied in detail it is difficult to summarise the directional accuracy of hearing in songbirds. However, it seems safe to conclude that the highest accuracy is about 20° in the horizontal plane, and that many species perform worse than this. We might have expected that birds which rely on songs and calls would be able to pinpoint each other with high accuracy. There are, however, two factors which probably explain how these birds cope with this relatively poor performance. First, they are highly mobile, and second, songs and calls are typically repeated many times in succession. Thus, by using short flights it is possible for a bird to home in on a sound source through successive approximations until visual contact with the singer or caller is possible. While these birds may not initially know with high accuracy the direction from which a sound is coming, they can soon find it, especially if the sound is repeated many times or given continuously.

Sound direction accuracy: owls

Not all birds are as poor as songbirds in determining sound direction; owls (Strigidae) and barn owls (Tytonidae) are notable exceptions. Both of these groups employ accurate sound location to guide their foraging, primarily using the sounds that prey animals make as they move through leaf litter or grass. While these sounds are not stereotyped and repeated in the way that songs and calls are, the rustles are produced for relatively long periods (at least a few seconds) at a time, and this allows an owl to locate and even track the movements of its prey. The rustles also contain a wide range of sound frequencies, and this probably also aids their detection and location.

Fifty years ago, a detailed study was published showing that Barn Owls *Tyto alba* were capable of capturing mice in total darkness, guided only by sound cues given out as the mice moved through leaf litter. It was shown that a Barn Owl could drop down from a low perch directly onto its prey, but if it missed, it could then capture the mouse as it moved away, generating more leaf-litter rustles. However, the owls could not complete these actions without practice. They had to have a period of a few weeks learning about the test situation, which was designed to mimic natural perches and leaf litter. There have now been studies showing this kind of behaviour in other owl species.

A number of studies have quantified the accuracy of owls' sound location abilities, and also determined the mechanisms that underpin it. These studies have used an owl's natural tendency to face towards a sound source that is suddenly presented. By carefully tracking owl head movements when the bird was presented with sounds in different locations, it was possible to determine just how accurately sound direction can be determined. Using bursts of broadband noise (to humans this sounds like a hiss, as it contains a wide range of frequencies) that are more than one second long, Barn Owls were shown to be able to locate sounds to an accuracy of 3–4°. This degree of accuracy is similar to what humans can achieve when tested under similar conditions.

FIGURE 6.5 The elaborate outer ear structures of owls. This side-on view of the head of a Long-eared Owl *Asio otus* shows the prominent 'long-ear' feathers after which the species is named. However, these tufts of feather have no association with hearing, and are used for visual signalling. The curving line of feathers at the side of head indicates the edge of the facial-disc feathers. When they are parted, as in the illustrations to the right, they reveal large ear openings, and it is possible to see into the ear canals. These openings are bordered by flaps of skin which hold small densely packed feathers. The ear openings and flaps of skin on each side of the head are of different size, shape, and vertical position. Although these differences are subtle, they play a crucial role in the ability of owls to locate sounds with high accuracy. When looking into the ears it is also possible to see the sides of the eyes, which bulge from the skull (Figure 4.2). See also Figure 5.15, which shows the flaps of skin in front of the ear openings in a Barn Owl. (Photo by Tom Koerner, USFWS, public domain; drawings by Rolf Åke Norberg.)

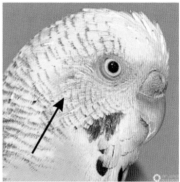

FIGURE 6.6 The ear openings of birds other than owls (Figure 6.5) are simple holes leading into the ear canals. They are positioned symmetrically each side of the head, behind and slightly below the eyes. In most bird species the ear openings are not visible because they are covered by a small tract of feathers, the ear coverts, as shown in this Budgerigar *Melopsittacus undulatus*. However, in some birds which have 'bald' heads, such as this Black Vulture *Coragyps atratus*, the ear openings can be clearly seen. (Photo of Black Vulture by Bryan William Jones, Budgerigar by Ian Junor.)

The key to the owls' directional accuracy lies not in the ears themselves but in the fact that owls have elaborate outer ear structures (Figure 6.5). In all other bird taxa, there are no such structures, and the entrances to the ear canals are just simple holes positioned below and behind the eyes. In most bird species these cannot be seen as they are covered by feathers (Figure 6.6). The ear entrances, and indeed the outer ear structures, of owls are also not seen since they are covered by the feathers of the facial disc. These are the tracts of feathers that give owls their distinctive 'owl-like' appearance. Despite their skulls having basically the same general shape as those of other birds, owl heads appear much broader thanks to the presence of their outer ears and the feathers that cover them.

These outer ear structures increase the effective distance between the ear openings of owls. They take the form of flaps of skin that are placed both before and behind the openings of the ear canals. Attached to these skin flaps are specialised small hard feathers which are capable of reflecting sounds. These are quite unlike the feathers that cover the ear openings in most birds, which are more or less transparent to sound. The positions of these flaps of skin, and hence the feathers, are under the bird's control, and changes in their position can result in dramatic alterations to the appearance of an owl's head, from narrow and vertically elongated to round and broad.

A further elaboration of these outer ear structures is that in most owl species the outer ears are asymmetric, differing in both size and position. We are used to the idea that our ears are more or less mirror images of each other. But in owls they are distorted mirror images. Basically, the ear flaps may be both higher-placed and larger on one side of the head compared with the other. Analysis of how these outer ear structures distort sounds before they enter the ear canals has led to a detailed understanding of how they enhance the accuracy of an owl's ability to determine the position of a sound. In humans our outer ears also distort sound before they enter the ear canals, and these are recognised as an important part of the mechanism that allows us to locate sounds vertically, as well as horizontally. Try gently distorting your own outer ears and you will find that sound location accuracy is compromised.

It has been shown that in owls the outer ears and their asymmetry have a key role in locating sounds in the vertical plane as well as in the horizontal plane. The result is that an owl can accurately determine the direction of a sound source within a large sector of the space that lies in front of the head. An owl does not have the rather vague idea of source direction that a songbird has, it knows rather exactly where a sound source is. It is this accuracy that provides a foundation upon which an owl can learn to capture prey in total darkness. The word 'learn' here is important. As described above, an owl needs experience within an experimental setup before it will exhibit the behaviour of catching prey using sound cues alone. It does not have to be explicitly taught how to do this, but it does need experience in the test situation. The way in which hearing is combined with experience in a particular habitat to complement vision in the nocturnal foraging of owls is discussed in detail in Chapter 9.

Sound ranging: determining the distance to a sound source

Knowing the distance to a bird that is singing nearby can be vitally important to another bird. For example, knowing this distance can indicate whether a potential mate is close enough to pursue, or whether a potential rival has crossed a territorial boundary.

As with sound direction, the ability to determine sound range has been measured in greatest detail in passerine species. It has not been found to be impressive. In most studies the test situations used have evoked territorial behaviour in response to the playback of a song of the same species. This confines most studies of sound ranging to males because it is usually male territory holders that will approach the source of the playback (typically a small loudspeaker) in an attempt to find a presumed rival. The distance at which a territory holder will start to approach or sing back against a presumed rival is used as a measure of the bird's sound ranging abilities.

Manipulation of the playback sound allows the identification of cues that a bird uses to estimate the distance to the sound source. This approach has shown that it is the volume (amplitude) of a familiar sound that is the key sound ranging cue used by birds: the quieter a familiar sound, the further away it is estimated to be. However, in natural situations sound volume can be an unreliable cue to distance, and this adds further to the inaccuracy of sound ranging.

Atmospheric conditions can change sound transmission significantly. Furthermore, even naturally produced bird sounds can vary in amplitude depending on the direction that the bird is facing. This is because the mouth acts as a mechanism that projects sounds away from the body, and the direction in which the calling or singing bird faces can influence sound volume. Some birds may scan their head around while singing, and this gives the impression of a sound source whose range is changing. This ventriloquial effect can be quite significant and often catches out a birdwatcher trying to locate a bird singing from vegetation.

The overall conclusion from studies of sound ranging is that it is primarily a cognitive ability. It depends not upon the specific hearing ability of the listening bird, but upon the bird's knowledge of the sound's original frequency spectrum and intensity. This means that a bird has to learn a lot about a specific song (or call) before it can determine the possible range of that song. This is paralleled by our own experience of learning bird songs and calls. It is difficult to learn songs in a short training session; it needs repeated exposure to the sounds given by a species in many different situations before we can be confident of which species is making the sound and how far away it might be. In short, accurate sound ranging depends crucially upon the sounds being familiar to the listener; that is, it must be heard in many different situations.

In support of this reliance upon familiar sounds for accurate ranging in birds are a number of field studies. These have all demonstrated that the degree of familiarity with a specific song type will strongly influence a male's ability to discriminate between degraded and un-degraded playback songs, and its ability to assess the distance of the sound source. There are, however, some field studies that have suggested that the distance to unfamiliar sounds can be effectively determined in certain situations. How rapidly an unfamiliar sound can become familiar, and how rapidly its patterns of degradation with distance and environmental types can be learned, may be crucial in the sound ranging ability of particular species.

Even with experience, the sound ranging ability of songbirds is surprisingly poor. A territory-holding male's response to the range of a sound is probably categorical. For example, the bird may sort sound ranges into just two categories, 'near' and 'far', rather than exhibit more nuanced judgements. A good example of this has been described in Common Chaffinches *Fringilla coelebs* (Figure 6.7). Their territorial songs and calls are heard frequently in a range of habitats across much of Europe during the breeding season, and these songs and calls have been shown to play an important role in the territorial spacing of breeding birds. Chaffinches, however, show only a categorical response to the playback of songs presented within a range of distances. The birds seem to distinguish sounds as being in only two categories: 'near' sounds (0, 20, and 40 m away) and 'far' sounds (80 and 120 m away). This range categorisation, added to the inaccuracy of location, results in a high degree of imprecision when determining the position of another bird singing or calling in the locality (Figure 6.7)

In the context of territorial defence this kind of categorical sound ranging seems to be all that is required. A resident male may simply categorise the songs of nearby males as potentially threatening when 'near', as it is likely to be inside its territory, or of no threat when 'far', which is likely to be outside its territory.

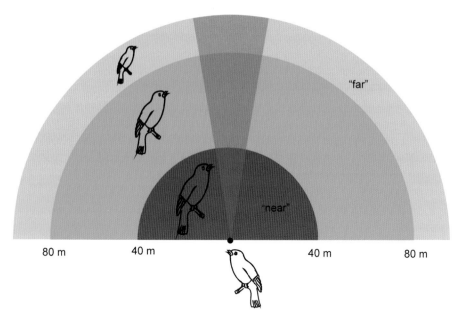

FIGURE 6.7 The imprecision of sound location and ranging in songbirds. This schematic diagram indicates how on the horizontal plane a bird may determine the position of another when hearing its song or call, and how it may respond to it. Directional accuracy of sound location is low. The listening bird (bottom) can determine the direction of a sound source with only low accuracy, with the degree of imprecision indicated by the grey shading – which in this example is a sector 20° wide. A singing bird may be positioned exactly in the centre of the grey area, but the listener can determine only that it lies somewhere within that grey zone. In addition, the listener cannot determine the distance or range of the singing bird with high accuracy. The listener may actually be able to determine only that the singer is 'near', in this case within 40 m (within the red zone) or 'far', in this case beyond 80 m (within the blue zone and beyond). If the singing is judged to be near, it may be a territorial intruder and therefore in need of investigation. If the singer is judged to be far, then it is probably regarded by the listener as of no threat or interest. If the singing bird is judged to lie in the intermediate (green) zone, the listener may well wait to hear the song or call repeated before responding. If the singer is moving the listener may wait until the sound can be judged to lie within either the near or far ranges.

There is evidence that Great Tits may be able to categorise sounds as coming from several different distances, but nevertheless their response may still be categorical, thus showing a rather crude determination of a sound's range.

Given the key role that sounds play in regulating the behaviour of songbirds, it may seem surprising that they are so poor at determining both the direction and range of a sound source. However, the key factor in the use of sound by these birds is that the sounds themselves are stereotypical for the species. This means that their characteristics in different environments and at different ranges can be readily learned. This, coupled with the high mobility of birds, enables them to quickly home in on a repeated sound through successive, if not very precise, approximations.

Even in owls, which have a more refined ability to determine the direction of sounds, the idea that an individual needs to be familiar with the characteristics of a sound source still seems to apply. Observations clearly show that an owl which pounces on prey in total darkness does indeed have reasonably accurate information on the distance to the sound. However, this is achieved over only short distances, a few metres, the range of a pounce from a perch onto the floor below. Furthermore, to achieve prey capture using sound cues, an owl has to have a high degree of familiarity with the experimental setup or natural situation. Thus, the birds need to have had a lot of exposure to the sounds of leaf-litter rustles caused by prey. In the original Barn Owl experimental studies, it took birds two to three weeks of learning to pounce on prey in reduced light, before doing so in darkness.

It is possible to train an owl to hunt for food items hidden in grass using a buried sound source. Some kind of clicker or white noise sound, triggered remotely, can be used. Birds trained in this way can provide interesting footage for wildlife films, as it seems to replicate what birds do in the wild. However, this technique requires the bird to be very familiar with the sound source and to receive regular training with it, in a range of circumstances. This again indicates that there is a significant cognitive component to sound ranging.

Active sonar: echolocation

Locating a sound source is sometimes described as *passive sonar*. Sonar is an acronym for so(und) na(vigation) r(anging). It is passive because it relies on the object of interest giving out a sound before its direction and range can be determined. There is also, however, *active sonar*, in which the location of an object is determined by sending out a sound signal to actively locate objects in the vicinity. It depends upon detecting and interpreting sound echoes reflected back from an object, hence it is also referred to as *echolocation*. Echolocation is usually associated with mammals and is found in a number of taxa which are not closely related, suggesting that it has evolved independently a number of times. However, echolocation is also found in some bird species, which employ it in quite specific situations.

As in sound ranging, active sonar depends upon knowledge of the characteristics of the source sound and knowledge of how this sound's reflections are modified by surfaces and objects in the environment. Humans are certainly capable of using active sonar for the location of objects, and this was first subject to detailed study more than 60 years ago. For humans, echolocation requires concentrated learning about sounds and their echoes in different situations. A visually impaired person regularly tapping a stick, or clicking their fingers, is sending out standardised pulses of sound. It is the echoes of these sounds that provide information about the environment. This ability is not special to visually impaired people, and it has been shown that these same techniques can be employed by fully sighted people. However, constant use and practice seem to be key in order to extract accurate information from the environment.

The key component of echolocation is the time delay between the sending out of each standardised sound and receiving its echo from an object. Time delay translates directly into range. Different surfaces also alter the nature of the echo and hence can provide information on the nature of an object. Finally, the nature of the sound emitted, especially its frequency, sets a limit on the smallest object that can be detected. The length of each sound pulse and the rate of repetition also affect the accuracy of the information that can be extracted from the echoes. Most of these limitations are determined by the physics of sound, not the hearing of the listener.

Mammal species which use active sonar to guide key behaviours include bats, toothed whales (cetaceans, Odontoceti), tenrecs (Tenrecidae) and shrews (Soricidae). In the bats and cetaceans active sonar functions to provide particularly fine spatial information about the nature, positions, and movements of objects in the environment. To achieve the extraction of fine spatial detail, sounds in the ultrasound range are used. They typically have frequencies up to 100 kHz, and most of the energy is in the range 30–70 kHz; such sounds are well outside our own hearing range.

Bats and cetaceans use these ultrasounds because it is only high-frequency sounds that can provide coherent reflections from objects of small dimensions; low-frequency sounds are usually not reflected from small objects; they simply pass around them or are scattered incoherently. However, not all sounds used by mammals to echolocate are ultrasounds. Some species of Old World fruit bats, tenrecs, and shrews, for example, use echolocatory sounds within the higher human audible range. As a result, these species cannot detect such fine spatial details as microbats.

The hearing of birds is within an even lower range of frequencies, so the size and the position of objects which can be located by active sonar in birds must be rather large. The physics of these sounds is against the extraction of high spatial detail. It is not surprising therefore that echolocation in birds is restricted to just a few species and used only in specific locations. However, in these situations active sonar provides key information.

Oilbirds *Steatornis caripensis* (Steatornithidae, order Caprimulgiformes) and cave swiftlets (Apodidae, order Apodiformes) are the only birds in which active sonar has been demonstrated (Figure 6.8). These two taxa are now regarded as closely related. The swiftlets have been subject to extensive recent taxonomic revisions and the total number of species is disputed. Oilbirds and swiftlets have in common the use of large caves for nesting and roosting. Typically, they enter deep inside these caves, into regions of complete darkness. It is thought that echolocation is used by the birds only when inside the caves.

Oilbirds are a species of large nocturnal nightjar which are found in northern South America and on Trinidad in the Caribbean. The echolocating swiftlets are a group of up to about 30 species (taxonomically they have proved difficult to sort out), which inhabit tropical regions from the Indian Ocean, through the Far East,

FIGURE 6.8 Birds that use active sonar. On the left, a Black-nest Swiftlet *Aerodramus maximus*, photographed in the Malaya Peninsula. On the right, two Oilbirds *Steatornis caripensis* roosting on a ledge in daytime in Ecuador. Both species exploit the resources of tropical rainforests; the Oilbirds consume fruits while the swiftlets are insectivores. (Photo of Swiftlet by wokoti [CC BY-SA 2.0], Oilbirds by Eric Gropp [CC BY 2.0].)

northern Australia, and the western Pacific Ocean. Unlike the Oilbirds they are not nocturnal, but they do fly regularly at dusk and dawn, and they may plunge into the darkness of caves directly from bright daylight.

The echolocatory abilities of both swiftlets and Oilbirds were first investigated systematically more than 60 years ago, but information is not available on the abilities of all swiftlet species due to the difficulty of working with them. They are difficult, if not impossible, to keep in captivity for long periods and furthermore they are unlikely to show natural behaviours requiring the use of their active sonar in captivity. Unlike the work with bats and cetaceans, this means that definitive studies of the echolocation of these birds have not been possible. However, some limited but robust data are available.

First, it is clear that vocal sounds produced by these birds when echolocating are quite distinctive. As in many of the echolocating mammals, the echolocatory sounds of birds are sharp single or double clicks, or trains of clicks, i.e. each sound is short with a clear start and finish. Unlike the sounds produced by echolocating mammals, they are at low frequencies, well within the range of our hearing, with most of the energy between 1.5 and 2.5 kHz in Oilbirds. In swiftlets, clicks were reported to be repeated with increasing rapidity when the bird was approaching an obstacle or entering a cave. This is much in the same way that bats increase their click rate when approaching an object. However, not all studies have recorded this in swiftlets.

Second, the smallest objects which a number of species of swiftlets could avoid were wooden poles with diameters of between 4 and 10 mm, while the smallest objects that White-rumped Swiftlets *Aerodramus spodiopygius* were reported to avoid were between 10 and 20 mm in diameter. These performances are not impressive compared with what bats can achieve in comparable tests, in which wires as small as 0.1 mm diameter can be detected. The minimum sizes of objects that Oilbirds can detect using active sonar are even larger than those detected by the swiftlets. Plastic discs of various diameter were hung in the passageway leading out of a nesting cave, and it was reported that all birds hit 5 and 10 cm diameter discs as if nothing had existed in their paths. The first signs of avoidance appeared when 20 cm discs were presented, and all birds avoided 40 cm discs. Another study, however, reported that Oilbirds could detect obstacles as small as 3.2 cm in diameter.

Third, the echolocatory sound pulses (clicks) produced by Oilbirds are very loud. Sound levels as high as 100 dB were recorded at a distance of 1 m from the birds, with the clicks produced up to 12 times per second. Certainly, the interiors of Oilbird caves are very noisy with large numbers of birds simultaneously and continuously producing streams of loud clicks when birds are preparing to leave the cave for their nocturnal foraging forays.

The performance of all these birds in terms of the threshold size of objects which can be detected would seem surprisingly poor, but they are well within what is predicted by the relatively low frequencies of the sounds that are emitted. However, the results of all the testing, plus field observations, indicate that active sonar can be used to guide these birds with respect to large objects. These include other birds and cave walls. Clearly, such low spatial resolution is sufficient to guide the flight and landing of Oilbirds and swiftlets in complete darkness.

Oilbird roosting caves are typically very noisy, and it seems likely that specific nest sites, which are on ledges on the cave walls, may be located using passive sonar, with flying birds using the calls of mates or young on nest ledges to guide them. Oilbirds and swifts seem to live highly predictable lives within the darkness of their roosting and nest caves, and Oilbirds may use the same nest ledge throughout their relatively long lives. This gives the birds plenty of time to learn how sounds are degraded, and how their echoes can be used to provide spatial information in complete darkness.

Summary: hearing in birds

There is an ever-growing body of evidence that sounds are a key source of information for birds. Songs, calls, and instrumental sounds provide information that regulates the everyday social behaviour of birds. Researchers who study bird songs and calls have provided compelling evidence that these sounds have been shaped through natural selection to communicate details about the species, sex, and physiological state of individuals.

All of these acoustic signals employ a relatively narrow range of sound frequencies. Most songs and calls are within the frequency range 1–5 kHz, which is audible to humans (at least to younger humans). The sounds that birds rely on are therefore within the hearing of humans; there are no 'secret' sound channels used by birds that humans cannot detect. Furthermore, because sensitivity to sounds in birds is similar to our own, or slightly lower, we should be able to hear the sounds produced by a bird at about the same distance as other birds in the locality.

Most bird species have a relatively poor ability to determine the direction and distance of a sound source. Even among songbirds, the accuracy with which they can identify the position of a sound source is not very great and is inferior to that of humans. Indeed, the range of a sound source may be determined only in broad categories. The actual position of a singing bird may be determined by approximations and approaches in a series of successive homing movements.

The direction and distance of a sound source can be determined with high accuracy by owl species. Even in these birds, however, sound range can be determined accurately over only a few metres. Furthermore, to be accurately located, the sound needs particular characteristics: it must have a wide range of frequencies, and it must continue for a few seconds or be frequently repeated. These are the sound characteristics of leaf litter or grass disturbed by an animal moving through it, and so it seems that the sound localisation of owls is adapted to the specific task of prey detection and capture.

In a handful of bird species active sonar underpins their use of deep caves as safe roosting and nesting sites. The resolution of this sonar is low and probably requires the birds to be highly familiar with a specific location.

Olfaction: information from smells

I first started research work on the senses of birds 50 years ago. At that time, the sense of smell in birds was considered something rather special. It was presented as a sensory capacity found only in a few species, serving a specific role in their foraging. However, the intervening years have seen something of a minor revolution. Authors now write about the excellent sense of smell of birds, its wide distribution across taxa, and its importance as a source of many different types of information.

Detailed knowledge of olfaction in birds remains, however, rather piecemeal. There is still much to learn about the tuning of olfactory abilities in different species, the behaviours in which olfactory information plays a key role, and the specific functions of identified chemical compounds in the social life of birds, as well as in their foraging. However, the pieces of information are numerous and far spread taxonomically. They indicate that olfactory information in many bird species is just as important for individual survival and reproductive success as the information derived from the other telereceptive senses of vision and hearing. It is not possible to say, however, which specific olfactory stimuli birds are gaining their olfactory information from. While we have a good idea of what birds use olfactory information for, it is often not possible to say exactly what they are responding to.

The olfactory systems of birds

The sense of smell is based on sampling air taken in through the nostrils. With each inhalation a sample of air is obtained, and each sample potentially contains a large number of different chemical compounds. The olfactory system analyses each sample of air for the presence and concentration of certain compounds, and if they are detected a signal is sent to the brain. Olfaction does not involve a complete chemical analysis of the sampled air; only certain compounds are detected by specific receptors.

In the majority of birds, the paired nostrils are positioned towards the base of the upper bill (Figure 6.9). They are always open, taking in a fresh sample of air on each inhalation. In some birds the nostrils can be closed. In gannets and cormorants (Suliformes), the nostrils open inside the mouth and so closing the mouth prevents water being forced into the nasal chambers when a bird enters water at high speed. In the surface-diving auks (Alcidae) the nostrils are narrow slits low down just above the cutting edge of the bill close to the mouth hinge; they can be closed by a flap of skin. In kiwi species (Apterygidae) the nostrils are positioned just behind the tip of the maxilla with their openings projecting laterally. Kiwi probe with their bills in leaf litter and soft substrates and they can be heard sniffing (or expelling air to remove material blocking the nostrils) as they forage on a forest floor.

In all bird species inhaled air passes through the nostrils and successively through three chambers (conchae). These chambers filter out small debris, warm and moisten the air, and finally sample it for chemical compounds. The third concha is lined with mucosa containing densely packed neural fibres. These fibres are the olfactory receptors, and they detect the presence of specific chemical compounds as the air swirls around the chamber before passing into the respiratory system. From these receptors neurons reach the brain at the olfactory bulbs and signal the presence of specific compounds in the inhaled air.

Olfactory bulbs are usually paired structures (sometimes they are fused) positioned centrally together at the front lower part of the brain (Figure 6.10). A lot of anatomical work has focused on olfactory bulbs because they vary markedly across bird species in both shape and size, and their dimensions can be readily measured and compared.

To date, the smallest reported olfactory bulb has a volume of just 0.06 mm^3 in Spotted Pardalotes *Pardalotus punctatus* (Passeriformes), while the largest has a volume of 217.63 mm^3 and is found in Emus *Dromaius novaehollandiae* (Casuariiformes). This is a 3500-fold difference in size, and so it is tempting to speculate that information extracted from smell must play a more important role in Emus than in passerines. However, experimental evidence that this is the case does not exist, although it has not been looked for. Furthermore, comparing olfactory bulbs is not the same as comparing the function and structure of the olfactory organs themselves. It is equivalent to comparing the areas of the brain devoted to the analysis of visual information, rather than comparing and understanding the structures of eyes and how they gather and constrain information.

FIGURE 6.9 The location of the nostrils is indicated in these photographs by white arrows. Among the water and diving birds of the Suliformes (cormorants, shags, gannets, and boobies) nostrils open inside the mouth, illustrated here by the example of a Great Cormorant *Phalacrocorax carbo* (top left). In the majority of birds nostril openings are in the maxilla (upper bill) and may take various shapes, sizes, and positions, illustrated here by a Black Vulture *Coragyps atratus* (top right), which has elongated slits, and by a Budgerigar *Melopsittacus undulatus* (middle right), whose nostrils are two small circular apertures at the base of the bill. In auks, illustrated here by an Atlantic Puffin *Fratercula arctica* (bottom right), the nostrils open at the lower edge of the maxilla close to the base of the bill. In tube-nosed birds (Procellariiformes), illustrated here by a Leach's Storm Petrel *Oceanodroma leucorhoa* (lower left), the nostrils open at the end of prominent tubes that lie along the upper edge of the maxilla. (Photo of Vulture by Bryan William Jones, Budgerigar by Ian Junor, Petrel by Sue Freiburger.)

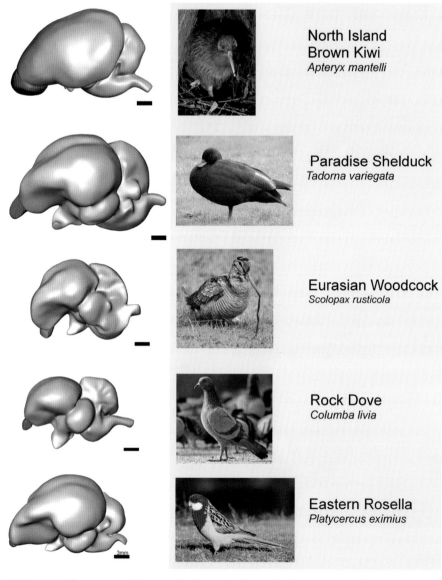

FIGURE 6.10 The sizes and shapes of olfactory bulbs in birds vary markedly across species. This sample is of five species whose olfactory bulbs range from among the largest to the smallest described to date. Each brain illustration is to the same scale (the scale bar is 3 mm) and shows a model of the complete brains with the olfactory bulb highlighted in blue. In the Rosella the bulb can hardly be seen, while in the Kiwi it is very prominent. Brain illustrations by courtesy of Jeremy Corfield. (Photo of Kiwi by Maungatautari Ecological Island Trust [public domain], Shelduck by Michael Hamilton [CC BY-SA 2.0], Woodcock by Ronald Slabke [CC BY-SA 3.0], Dove by Muhammad Mahdi Karim, www.micro2macro.net [GNU free documentation licence], Rosella by J. J. Harrison [CC BY-SA 3.0].)

Evidence for the importance of olfaction in birds

The first behavioural studies of olfaction were carried out in kiwi species. They were chosen because they were known to sniff, had nostrils at the tips of their bills, and had large olfactory bulbs. These studies, published 50 years ago, showed that kiwi use olfaction for food location. This was followed 20 years later by experiments which demonstrated that New World vultures (Cathartidae) could also find hidden food using smell, in this case carrion. Another 10 years passed before a specific role of olfaction in food location in petrels (Procellariidae) was demonstrated. During this century there have been a number of studies showing that olfaction can play a key role in behaviours other than foraging, including the regulation of social behaviour and species recognition.

This range of bird species in which behavioural evidence has demonstrated a specific function of the sense of smell is quite impressive. It includes bird species which have both relatively large and relatively small olfactory bulbs. The latter point is particularly important because it shows that although a greater proportion of the brain is devoted to the analysis of smell in some birds, the presence of a small olfactory bulb does not mean that smell is non-functional or of little importance. Recent analysis of olfactory bulb size in a sample of 135 bird species concluded that the absolute size and proportional size of olfactory bulbs is complex and does not show a simple relationship with foraging behaviour and habitat. However, these comparisons do support the conclusion that olfaction is probably an important sense in all bird species, although its specific roles in most species have yet to be understood.

Uses of olfactory information

The first concrete evidence of how olfactory information is used by birds focused upon foraging. However, even in foraging, olfactory information has been shown to be used in two distinctive ways: finding specific items and finding good foraging locations.

Finding specific items

The best examples of the use of olfactory cues in foraging for individual prey items come from various species of kiwi, and from New World vultures, with olfactory foraging being especially important in Turkey Vultures (Figure 6.11). The phylogeny, behaviour, and ecology of these birds are quite different, but it has been shown that olfaction plays a similar key role in finding hidden food items in both taxa.

Olfaction is used by kiwi in the detection of soil invertebrates, and by New World vultures to detect decomposing carcasses. In both species olfactory cues are important in finding food items that are usually hidden from sight. In kiwi, their key food items, earthworms, are buried in the soil. In the vultures, the use of olfaction is particularly valuable when carcasses are hidden under vegetation. Evidence that other birds use olfaction to find individual food items is lacking, but it has probably not been sought.

FIGURE 6.11 Two bird species in which olfactory cues play a key role in locating individual hidden food items: North Island Brown Kiwi *Apteryx mantelli* and Turkey Vulture *Cathartes aura*. (Photo of Kiwi by Maungatautari Ecological Island Trust [public domain], Turkey Vulture by Dori [CC BY-SA 2.0].)

Finding foraging locations

Many birds forage on food items which may be widely dispersed in the environment, but when they do occur they can do so in substantial concentrations. These concentrations constitute highly profitable foraging areas. However, finding these areas may not be straightforward, especially if the habitat is somewhat uniform. Studies of pelagic seabirds provided the first, and a highly compelling, example of how olfaction is used to find these profitable prey patches.

It is the presence of one particular chemical compound, dimethyl sulphide (DMS), that is key to this. This compound does not directly indicate the presence of the birds' preferred prey items, but rather it is a surrogate. DMS is produced by marine phytoplankton when it is grazed by zooplankton. It is often described as 'the smell of the seaside' and used to be commonly referred to as the smell of ozone, but ozone does not have a smell. When DMS occurs at high concentrations in the ocean it is associated with areas of high primary productivity. In turn, these are areas where the larger animals, which exploit the food chain that is built upon this high productivity, concentrate. It is these animals which form the preferred prey of particular seabirds. Pioneering work has shown that the presence at high concentrations of DMS are indeed detected and exploited by various petrel species (Procellariiformes), and large numbers of these birds may be drawn upwind to a DMS concentration. Furthermore, they exhibit particular flight patterns, zigzagging across the ocean, aimed at detecting plumes of DMS; once a plume has been detected the birds fly upwind towards the source (Figure 6.12).

A similar use of odour concentrations to locate a profitable foraging area has been demonstrated recently in the terrestrial environment. Great Tits forage mainly in woodland habitats, and it has been experimentally demonstrated that they may be attracted to concentrations of odours that are produced by tree foliage following its attack by herbivores, particularly the caterpillars of moths. These caterpillars are a preferred food of these birds, especially for the provisioning of young. Thus, the chemicals that are released by the attacked plants seem to provide a signal which indicates that there is a local concentration of the Great Tits' preferred food (Figure 6.12).

Semiochemicals: information from body odours

The chemical compounds exploited by foraging birds are by-products of other processes. It would be difficult to argue that they are emitted for a specific purpose. Semiochemicals, on the other hand, are substances produced by an organism for the specific purpose of communication between individuals of the same species.

DMS emitted when the phytoplankton is grazed upon by zooplankton

birds fly upwind using a lateral and vertical zig-zag path to aid detection of the DMS plume

sea surface

phytoplankton

zooplankton

FIGURE 6.12 Two bird species in which olfactory cues play a key role in finding locations where food items may be abundant and hence foraging is profitable. At certain times of the year, as seasonal tree leaf growth gets under way, Great Tits *Parus major* may use odour cues to locate trees in which leaf-eating insects are abundant. Antarctic Prions *Pachyptila desolata* use odour cues to find patches of ocean where small prey items may be abundant in surface waters. The techniques that the prions use when searching the open ocean are well known and involve flying upwind using zigzag flight paths to intercept odour plumes. Once a plume has been detected the birds fly upwind until they find the source. The odour detected is dimethyl sulphide (DMS), which is released as a result of zooplankton consuming phytoplankton close to the surface. The abundant zooplankton attracts small predators which are the food source that the prions are searching for. The specific odours that Great Tits exploit are unknown, and it is not known if they use a specific flight pattern to detect and follow it. (Illustration of foraging prions modified from an original courtesy of Gabrielle Nevitt; photo of Antarctic Prion by Liam Quinn [CC BY-SA 2.0].)

The existence of semiochemicals was first established in mammals and insects, and they include the well-studied group of compounds known as pheromones. These are chemical substances capable of directly eliciting a social response in members of the same species, and they typically play a role in initiating and sustaining behaviours involved in reproduction. But not all semiochemicals are pheromones. Most semiochemicals are important in regulating social behaviour but they do not necessarily elicit a specific behavioural response. That odours produced by birds can function as semiochemicals has become clear only recently.

Odours and species recognition

The uropygial gland is the most important source of body odours in birds. This gland is situated low on the back, above the tail, and is also referred to as the oil gland. Its primary function is the productions of waxy fluids used as 'preen oils' to maintain the flexibility and waterproof properties of contour and flight feathers. A bird gathers oil from the gland in its bill and spreads it on its feathers. Any odour compounds produced by the uropygial gland will also be spread over feather surfaces and released into the air over a sustained period, much in the same way that perfumes are released from their oil base over a protracted period.

The secretions of uropygial glands vary markedly between species, and there is evidence that odours associated with these differences in chemical composition are detected by birds. However, evidence is often indirect because it is difficult to carry out fully controlled experiments to extract the key volatile compounds from the preen oil and then test them for a behavioural response.

Evidence of species recognition based on odours has been accumulated in a wide range of species. The first evidence came from studies of some petrel (procellariiform) species. Subsequently, behavioural evidence of species recognition has also been found in some passerines (Dark-eyed Juncos *Junco hyemalis*, waxwings *Bombycilla* spp.) and in Budgerigars *Melopsittacus undulatus*. Both species and sex recognition based on odour has been reported in Spotless Starlings *Sturnus unicolor*.

It seems that odour signals can differentiate between closely related species. For example, in shearwaters of the *Calonectris* genus it has been shown that body odours may strongly differ. It has also been argued that two species of waxbills (Estrildidae), Zebra Finches *Taeniopygia guttata* and Diamond Firetails *Stagonopleura guttata*, may also be differentiated by odour. However, behavioural evidence that these birds actually exploit this source of information is lacking.

Taken together, the above examples do suggest that there is a rich diversity of odour signals available in the preen-gland secretions of different bird species, and in some species they have been shown to function in species recognition. No doubt many more examples will soon be added to this list as researchers sample the uropygial secretions of more species, although the identity of the specific compounds used in species recognition has yet to be determined.

FIGURE 6.13 Three species of seabirds (Procellariiformes) which have been shown to use odour for individual recognition: Blue Petrel *Halobaena caerulea* (left), Antarctic Prion *Pachyptila desolata* (middle), Wilson's Storm Petrel *Oceanites oceanicus* (right). (Photos of Wilson's Storm Petrel and Blue Petrel by J. J. Harrison [CC BY-SA 3.0], Antarctic Prion by Liam Quinn [CC BY-SA 2.0].)

Odours and recognition of individuals

That birds can recognise individuals based upon odour cues alone has been demonstrated in a number of procellariiform seabird species. For example, the chicks of European Storm Petrels *Hydrobates pelagicus* are able to differentiate between individuals on odour alone. The same has been shown among adult Antarctic Prions, Wilson's Storm Petrels, and Blue Petrels (Figure 6.13). However, the source of these individual odour signals has yet to be found.

An interesting source of evidence that birds could potentially discriminate between individuals based on odours has come from experiments using mice as 'odour detectors'. Mice can be trained to differentiate between the odours of individual Chickens *Gallus gallus domesticus*. While this shows that Chickens do have individual odour signatures, it has yet to be shown that the birds themselves are able to detect this information.

Odours, mate choice and mating

Perhaps one of the most surprising recent examples of the role of odours in bird behaviour has come from studies of Crested Auklets (Figure 6.14). These birds produce an odour described by humans as a 'citrus-like' scent. Production of this odour is seasonal and is associated with a display behaviour, called a 'ruff-sniff', which involves a bird rubbing its face in the nape of a displaying partner. It is the ruff feathers which contain the odour, and it appears not to come from secretions of the uropygial gland but from feathers high up on the back. This secretion is thought to have a function as a parasite chemical repellent. It has been proposed that exchange of this repellent through ruff-sniffing can provide information on the health status of the individuals. Therefore the presence, and possibly the concentration, of these odours could be contributing to mate choice.

Secretions from the uropygial gland may play an important role in the control of mating in Chickens. It seems that cocks preferred to mate with hens whose

FIGURE 6.14 Crested Auklets *Aethia cristatella* emit a citrus-like scent from feathers around their necks. The prime function of the chemical may be to repel parasites, but its presence seems to be detected during display behaviours between pairs and may signal the health status of an individual. (Photo by F. Deines, USFWS [public domain].)

uropygial glands were intact as opposed to hens that had had the uropygial gland surgically removed. This difference disappeared in cocks that had had their olfactory bulbs removed and so had no sense of smell. This suggests that the behaviour of the cocks was based on the detection of odour cues. However, mating is a complex behaviour and just what is being detected by the cock from the odour is not clear; it could be recognition of the species, the sex, or the reproductive status of the female, or indeed all of these.

Odours and nests

That odours can play a role in nest building and nest maintenance has been demonstrated in a number of bird species. Some species of passerines, tits (Paridae) and starlings (Sturnidae), are well known for incorporating aromatic plant materials in their nests (Figure 6.15). Often these aromatic items are added during the lifetime of the nest, not just in the initial building, and it seems likely that the plant material helps to repel parasites. In turn this should help to enhance the growth rates of nestlings. That starlings use odour cues to select suitable material for their nests is well established. In tits the parents appear to use odour cues to determine when to replenish the nest with fresh aromatic herbs.

In addition to guiding these nest embellishments, there is good evidence that odours are used by some species to actually find their nests. It would be reasonable to assume that most birds should know exactly where their nests are located. However, like other petrels, Leach's Storm Petrels breed in dense colonies,

FIGURE 6.15 Common Starlings and Great Tits are among species which incorporate fresh aromatic plant materials in their nests. The materials are selected using odour cues. (Photo of Great Tit by Bengt Nyman [CC BY-SA 4.0], Common Starling by Pam Parsons, Pam P, via Flickr.)

sometimes containing many thousands of nests hidden in burrows. Furthermore, they return to their colonies and enter their nests only at night. Often, they avoid even moonlit nights because of their vulnerability to attack by predators. Thus, finding their nests by visual cues may not be possible. It has been shown that Leach's Storm Petrels use smell to track down their own nest burrow. To humans these nest burrows have a strong musky odour. It has been shown, using choice experiments involving nest material retrieved from different nests, that individual petrels are able to identify their own burrow by smell alone. The source of the musky odour and the individual nest signature, however, remain unknown.

Odours and navigation

It has often been speculated that seabirds could find their way to a breeding colony by flying upwind to the odour plume it emits. This certainly seems a possibility, but no definitive studies have shown it. Seabird colonies do emit strong odour plumes, strong enough to be detected by humans over a distance of many kilometres, and most birdwatchers soon come to know the smell of a seabird colony.

Experimental studies suggest that terrestrial species may also use smell-based navigation cues. Pigeons (Rock Doves) trained to home to a loft have been shown to learn odour cues associated with their home location. They use these odour cues to orient their homing flight when released from a remote location. This work was originally done in Italy, near the city of Pisa. It involved releasing groups of birds trained to fly to the lofts, some of which had their sense of smell disrupted at the time of release. Those treated in this way had difficulty in determining their homeward orientation.

The problem with this work was that it could not be replicated by other groups of researchers working in other locations. A consensus has emerged, however. It seems agreed that the original findings were soundly based and that birds in some locations may use olfactory cues to determine their homeward orientation. The explanation is that homing pigeons rely upon a number of different cues to guide their initial homeward orientation, and odours can be part of this depending upon whether there are consistent odour sources in a locality. If there are, then the birds may build them into the repertoire of cues that can guide them. One intriguing aspect of this work is that the actual odour is not fixed, but rather is something that is learned from the locality. It could be pollution from an industrial site or the smell of vineyards. As long as an odour can provide a reliable beacon, the pigeons may learn to use it.

Summary: olfaction in birds

Odours have been shown to provide many different types of information that are vital in the lives of birds of many species. This has been demonstrated in bird species from a wide range of orders and families, and species which differ in their behaviour and ecology. Odour information has been shown to play a part both in foraging for specific items and in the location of profitable foraging areas, and key roles in regulating social behaviour and reproduction. As with other senses, evidence for the roles and uses of information comes from a small sample of the total diversity of birds. This seems, however, to be primarily because the use of olfactory information has not been investigated, not because it does not occur. Because of this we may be only at the beginning of a full understanding of the roles that information provided by the sense of smell plays in the lives of birds.

Chapter 7

The intimate senses: touch and taste

The emergence of vision was a game changer in the evolution of animals. The ability to gain information about distant objects opened up many and diverse evolutionary pathways leading ultimately to the diversity of modern animals. However, the emergence of this *telereceptive* sense did not eclipse the importance of the *intimate* senses. These senses had evolved long before vision and provided information about objects and substances that were either in contact with, or even within, the body of an animal. In providing this information the intimate senses of touch and taste have remained vital for determining the daily wellbeing and survival of all animals. In fact, they capture the basic information that an animal must have in order to thrive.

It is tempting to think that in most birds the information provided by touch and taste is best considered as functioning in the background. We might think that touch and taste mainly help a bird to tick over rather than being responsible for providing information vital for the conduct of more exacting behaviours, such as foraging or the maintenance of social relationships. It is true that, in general, the senses of touch and taste do not play roles in behaviours that are considered glamorous, behaviours that grab our immediate attention. However, in some bird species, touch and taste have evolved to take a central place in controlling foraging. They provide information for locating food items every bit as important and subtle as the role played by vision, hearing, and olfaction in other species.

Our interests in the senses of vision, hearing, and olfaction arise partly because they are based on highly sophisticated organs, special anatomical structures in which sensory receptors are organised. With eyes, for example, it has been possible to compare and contrast their diversity to provide clues about their particular functions in different species. However, the receptors of touch and taste are typically widely scattered in the body, not grouped in clearly identified organs. Determining their specific roles in particular species is often a difficult task. However, in some bird species there are concentrations of receptors which together can be considered to function very much like organs. But even these concentrations of touch and taste receptors are not conspicuous and do not readily attract our attention. Thus, not only are these the intimate senses, but to understand them requires a more intimate knowledge of birds.

Touch

'Touch' is a complicated sense. Like vision and hearing it is multifaceted, with different components providing different types of information. When we casually speak of 'touch' we are in fact bundling together different information derived from separate types of receptors, each indicating a different property of an object that is in contact with our body. Much in the same way that we cannot readily separate the dimensions of acuity and colour vision in our everyday visual experiences, it is difficult to make ourselves aware of the different dimensions of touch.

We can readily appreciate, however, that there are marked differences in our sensitivity to touch across our bodies. These differences are readily explained by variations in the distributions of touch receptors across our body surfaces. It is a small step to appreciate that across species there can be marked differences in touch sensitivity and that different aspects of touch can vary in their importance in different species. Furthermore, it is possible to understand these interspecies differences in touch sensitivities in an ecological context, and to understand the part they play in the execution of particular tasks.

Because 'touch' is the amalgamation of several different types of information or sensations it is preferable, scientifically, not to refer to touch. Rather the term *somatic sensitivity* is used, and each type of sensitivity is described separately. Somatic sensitivities include pressure, skin stretch, vibration, noxious stimuli, and temperature. Each of these sensitivities involves different receptor types (Figure 7.1).

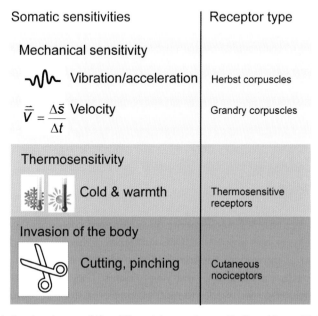

FIGURE 7.1 A simple schema of the different types of somatic (touch) sensitivities found in birds and the principal types of receptors which detect the different stimuli.

These somatic sensitivities are detected mainly through the surface of the skin. However, somatic receptors may also occur at locations within the body itself, and some aspects of touch may effectively reach a little outside the body. This is achieved by bristles and specialised feathers whose displacement is detected by receptors within the skin around the bristle and feather bases. All of the main somatic receptor types that are found in mammals have been reported in birds. This suggests that somatic receptors have a very long evolutionary history and have been little modified over time. What do seem to have been readily modified through natural selection are the distributions and relative numbers of different receptors in different species. This is much akin to the way the rod and cone photoreceptors differ little between bird species but their distributions and relative abundance across the retinas of different species vary markedly and provide very different visual capacities (Chapter 4).

While it seems possible that every bird species possesses all of the main types of somatic receptor, detailed studies are restricted to just a few species. Furthermore, quantitative studies on somatic sensitivities are very few. Therefore, much of what follows is based on rather broad generalisations. However, some intriguing examples of how birds employ information from somatic receptors will be discussed. It is clear that further research on the numbers, distributions, and sensitivities of somatic receptors across bird species could throw up some fascinating findings on the sensory ecology of touch.

Somatic sensitivities

Different somatic receptors in the skin detect mechanical, thermal and noxious stimuli coming from outside the body. Other somatic receptors provide information about internal conditions within the limbs, gut, cardiovascular and respiratory systems. Classifications of somatic receptors are usually based upon the stimuli that they respond to rather than on their detailed structure.

Most of the recent and detailed studies on the form and function of somatic receptors in birds have concentrated on those found in the bills of certain species. This is partly because they are readily identified and located but, more importantly, because their function has been shown to be key in the detection and identification of certain prey or food items.

Mechanoreception

Mechanoreceptors respond to the movements of an object or fluids with which they are in direct contact. Movements have different components: acceleration (vibration), velocity, and amplitude. Different types of mechanoreceptors respond to each aspect of movement separately. There are two main types of mechanoreceptors in birds, Herbst and Grandry corpuscles. They are not found in other animal taxa including mammals and reptiles. However, these receptors have close functional equivalents in other animals. In mammals they are referred to as lamellar type or Vater–Pacinian corpuscles.

Herbst corpuscles

Herbst corpuscles are named after the German anatomist Gustav Herbst, who first described them in birds' legs in 1848. They are the most widely distributed mechanoreceptors in a bird's body, occurring everywhere in the skin, along the large bones of the legs and wings, in tendons, muscles, and joint capsules, and near large blood vessels.

Herbst corpuscles respond to rapid pressure changes and are regarded primarily as vibration detectors. They can respond directly to a rapidly vibrating stimulus in a one-to-one manner (each movement eliciting a separate signal) at frequencies between 100 Hz (cycles per second) and 1000 Hz. However, while they signal the frequency of these vibrations, they do not detect either the amplitude or the velocity of the vibrating stimulus. The structure of Herbst corpuscles varies markedly depending on their location and with bird species, and so they are regarded as a class of receptor, rather than a single type.

It is in the bills of birds that the distribution and number of Herbst corpuscles show marked variations between species. High numbers and concentrations in particular locations along the bill are related to the way in which the bill itself is used as an exploratory device, especially during foraging. In those species of shorebirds (Scolopacidae) which probe with long bills in mud or sandy substrates when foraging, the tip of the bone of the upper jaw (the premaxilla) may bear countless small pits which are packed with Herbst corpuscles. These high concentrations of somatic receptors are known as *bill-tip organs*, and they sit just below the soft skin which covers the outer surface of the bill (the rhamphotheca) in these shorebird species. Thus, they function through the skin.

Concentrations of pits beneath the skin of the bill are also found in ibises and kiwi, but pits which open directly to the surface along the edges and inside the bill tip are found in the bills of ducks and geese, and parrots. These arrangements are described in greater detail in the 'Bill-tip organs' section below. In grain-feeding songbirds, especially the finches (Fringillidae), Herbst corpuscles and other types of mechanical receptors are located at places in the bill which are mechanically involved in seed-opening. In the long tongues of woodpeckers (Picidae), which are used for extracting invertebrates from narrow holes, Herbst corpuscles are also numerous.

In feathered skin, Herbst corpuscles are located primarily at the bases of the feather follicles and bristles. These have been interpreted as a mechanism which detects the vibrations of feathers and bristles. They probably provide birds with information on the stresses on feathers, particularly as feathers are moved by air currents during flight, and also enable birds to detect objects that touch the bristles. This function may apply particularly to bristles around the mouth (rictal bristles) and may enable adults to feed their young in the dark of a nest cavity.

Grandry corpuscles

These receptors primarily detect the velocity component of a moving object in touch with them. They are found particularly in feathered skin but also in the skin of the soft part of the bill covering of geese. Unlike Herbst corpuscles they do not follow vibrations and are thought to be sensitive mainly to velocity. Movements which cause crumpling of the skin seem to be the main type of stimulus that causes these receptors to respond.

Thermosensitive receptors

Humans are readily able to detect a source of heat or cold, particularly on certain of our body surfaces. For example, we seem to be particularly able to detect the radiant heat from the sun falling on our back or on our face. This ability is due to what are termed *thermosensitive units* which respond to both cold and warmth, and which can also respond to both the amplitude and rapidity of a temperature change. Although the presence of such receptors has been demonstrated in very few species of birds, they may be widely distributed among all bird species. They seem to play a role in the control of behaviours involving the regulation of body temperature, but are also probably involved in such behaviours as regulating nest temperature during incubation and chick rearing.

Cutaneous nociceptors

These somatosensory receptors are probably very important for the survival of all animals. They respond to invasion of the body, for example cutting or pinching. In mammals, their response seems to directly elicit significant changes in the condition of the body, such as an increase in blood pressure, heart rate, or respiratory frequency, when invasion is detected. Little is known about nociceptors in birds, but they have been identified in some species and they probably have similar properties to those found in the skin of mammals.

Bill-tip organs

Bill-tip organs were first described in geese and ducks. Recent research has also described other types of bill-tip organs in kiwi, parrots, long-billed shorebirds, and ibises. All bill-tip organs are high concentrations of somatic receptors, mainly Herbst corpuscles, arranged around the tips of the bill. The diverse groups of birds in which these organs occur, and the different arrangements of their receptors, indicate that bill-tip organs have evolved independently a number of times. In some species the bill-tip organ extends along the bill towards its base.

The importance of bill-tip organs is that they turn the bill into a self-guided exploratory device. The bill does not need visual guidance for it to detect food items. Bill-tip organs in parrots allow them to use their bills for manipulation of objects and as an extra 'limb' for climbing. The presence of a bill-tip organ means that vision can be more fully devoted to vigilance and predator detection. There is no need to have a binocular field which encircles the bill; hence it is birds with bill-tip organs that achieve the most comprehensive visual coverage of the world about them (Chapter 5).

The bill-tip organs of ducks and geese

Cursory inspection of the bills of geese and ducks gives no indication that they might be particularly sensitive to touch. The outer surface at the tip of the upper bill, and the outer and inner surfaces of the lower bill, are hard and horny. However, looking at the rim of the inside of the bill reveals, even to the naked eye, small pits. These pits contain touch papillae which reach the surface and transmit mechanical information to clusters of both Herbst and Grandry corpuscles, embedded in the hard horn of the bill tip (Figure 7.2).

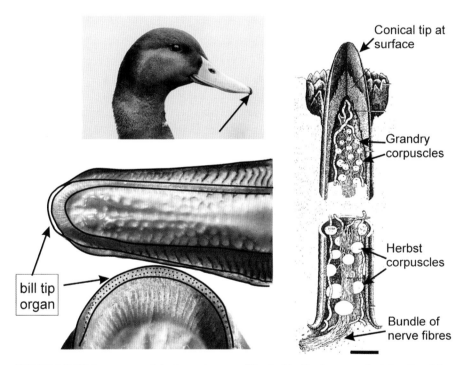

FIGURE 7.2 Bill-tip organs of ducks and geese. Situated in the horny nail at the tip of the upper bill (top), the organ appears as closely packed pits around the rim of the nail inside the mouth (lower diagrams). Each touch papilla (right) is embedded in the hard horn of the bill tip with just the conical tip reaching the surface. Mechanical movements of objects in contact with the conical tip of the papilla are transmitted down the papilla to be detected by clusters of Grandry and Herbst corpuscles. Signals from these receptors enter the nervous system via a bundle of nerve fibres. The scale bar in the diagram of the papilla is 100 μm. (Original drawings of bill-tip organ and touch papilla have been modified with permission of the original author, Herman Berkhoudt. Mallard photo by Bryan William Jones.)

The number, size, and shape of these clusters of receptors vary between wildfowl species. These differences probably indicate a fine-tuning of the structure of the bill-tip organ to meet the sensory demands of detecting and identifying various types of food items using touch cues alone. Neighbouring touch papillae are mechanically isolated one from another, and this suggests that fine spatial tactile discrimination by the bill-tip organ should be possible. For example, wildfowl may be able to discriminate between different objects and/or the surface structures of objects using tactile information alone. Indeed, there is experimental evidence that Mallard ducks can distinguish between real and model peas buried in soft substrates using only cues derived from the bill-tip organ.

The bill-tip organs of parrots

Bill-tip organs were first described in parrots by the French anatomist D. E. Goujon in 1869. Despite the general fascination with parrots, and their popularity as cage birds, it was more than a century before a further description of the bill-tip organ in a parrot was published. Even now, knowledge of parrot bill-tip organs remains rudimentary and lacks a comparative base.

The bill-tip organs of parrots are very different from those of wildfowl. Rather than continuous rows of touch receptors arranged around the bill tips, the receptors in parrots occur in groups of discrete bundles. These are well separated from each other in a symmetrical pattern along the inside edges of the highly curved upper mandible. In Senegal Parrots, for example, the bill-tip organ consists of seven pairs of clusters embedded in the hard horn of the bill with a single pit cluster at the bill tip (Figure 7.3).

The capabilities of such arrangements of touch receptors are not known in detail. Because of their spaced distribution they are unlikely to provide the kinds of fine-grained spatial information that are thought to be available through the bill-tip organs of wildfowl. In parrots, bill-tip organs may function in the manipulation of objects held in the bill tip. Parrots are notably good at this kind of manipulation with both hard and soft objects, such as sticks and soft fruits. It is also likely that the touch receptors are key in the positioning of the bill when it is used as a 'third limb' in climbing. Parrots typically reach out with their bill and pull themselves up using it, or gain extra stability when climbing head-downwards. This may be particularly important since their visual fields and bill shape preclude them from seeing their own bill or what is held in it.

The bill-tip organs of shorebirds, kiwi, and ibises

A third type of bill-tip organ occurs in shorebirds (Scolopacidae), kiwi (Apterygidae), and ibises (Threskiornithidae) (Figure 7.4). Because of the disparate evolutionary origins of these taxa (kiwi and shorebirds last shared a common ancestor 70 million years ago) it seems highly likely that the similarity of these bill-tip organs is due to evolutionary convergence. That is, they evolved separately as similar solutions to the sensory challenges associated with finding deeply buried prey items. Kiwi,

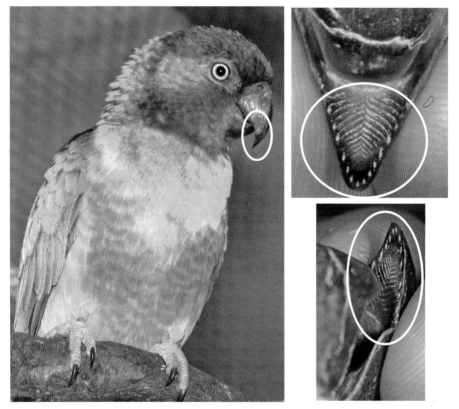

FIGURE 7.3 Bill-tip organs of parrots. Clusters of touch receptors in this Senegal Parrot *Poicephalus senegalus* are arranged in groups inside the upper bill in the curved tip part (left). This part of the bill is only keratin; there is no underlying bone. Inside the bill tip (right) clusters of touch receptors are symmetrically arranged just inside the edge of the bill, embedded in the horny keratin. The patterns of lines inside the bill are not part of the bill-tip organ and probably help to grip objects held in the bill tip or when the bill is used as a 'third limb'.

shorebirds, and ibises all probe with their long thin bills into soft soil and mud in search of invertebrate prey.

The function of these organs is best understood in shorebirds. It has been shown that their bill-tip organs are sufficient to locate prey and possibly to identify it. The bill-tip organ provides information on the vibrations produced as prey moves within muddy and sandy substrates. The bill tip does not have to make direct contact with a prey item for its presence to be detected. This is because as the bill is thrust into wet mud or sand, it displaces water away from it. If this displaced water encounters an object it creates a back pressure, and this can be detected by the organ. This ability has been given the name 'remote touch' because back-pressure patterns are produced within a cylindrical volume around the bill tip, perhaps up to 10 mm from the bill. Further probing may be necessary

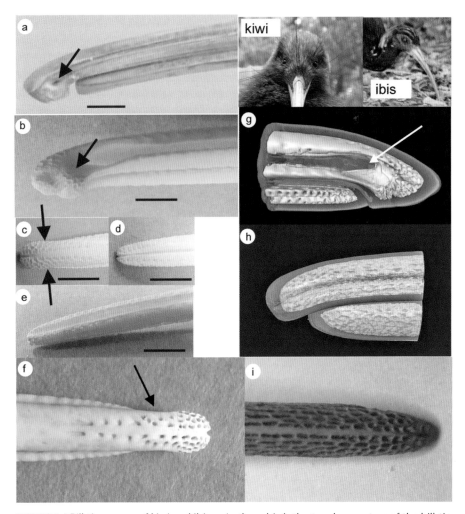

FIGURE 7.4 Bill-tip organs of kiwi and ibises. In these birds the touch receptors of the bill-tip organs are housed in clusters in pits in the bone beneath the keratin sheath that covers the bill. Photos *a* to *g* show the bill in Southern Brown Kiwi *Apteryx australis*. In *a* the keratin sheath is in place; the other photos show the pits in the bone of the upper bill (*b*, *c*, *f*, *g*) and lower bill (*d*, *e*, *g*). In all illustrations the arrows indicate the position of the nostrils (see Chapter 6). The pits in the upper and lower bills of a Madagascan Ibis *Lophotibis cristata* are shown in *h* and *i*. (Photos by the author and with permission from Susie Cunningham and Jeremy Corfield.)

to determine the exact position and identity of the object, but moving the bill around after inserting it may serve to locate the item, with the curvature of the bill helping to search a volume below the surface. The birds may detect a soft-bodied food item such as a worm, or a hard one such as a mollusc shell, and they may be able to tell the difference between them without making direct contact.

This type of bill-tip organ consists of clusters of pits within the bone around the tips of the maxilla and mandible, and in some species the pits are so crowded that the bone has a honeycomb-like appearance (Figure 7.4). The pits contain bundles of both Herbst and Grandry corpuscles. Unlike the bill-tip organs of wildfowl and parrots, however, the receptors are not exposed, but lie beneath the soft pliable skin that covers the bill. Thus, the movements of mechanical stimuli are sensed through the skin, much in the same way that most somatosensory receptors can pick up information through skin rather than being stimulated directly by objects.

Ibises have bill-tip organs like those of shorebirds and kiwi, and comparative studies have shown good evidence of a subtle tuning of the extent of their bill-tip organs in relation to differences in foraging tasks. Two parameters of ibis bill-tip organs – the length along the bill that sensory pits extend from the tip, and the total number of sensory pits in the bill – show a clear relationship with the type of habitat in which the birds probe.

Probing in more aquatic habitats is correlated with a greater number of sensory pits which extend further up the bill. Thus, more extensive bill-tip organs are found in Glossy Ibises *Plegadis falcinellus* while less extensive ones are found in Buff-necked Ibises *Theristicus caudatus*. Glossy Ibises forage almost exclusively in standing water, while Buff-necked Ibises are dry-land foragers. The explanation is that for the dry-land foragers prey items occur only near the bill tip, while an ibis foraging in water and softer substrates may encounter prey at a range of depths. Therefore, touch sensitivity further along the length of the bill will enable the location of prey at a range of depths.

Conclusion: the functions of touch sensitivity in birds

Somatosensory receptors primarily provide information about objects at the body surface. They may also provide information on slightly more remote events through detecting the displacement of feather and bristles. Such types of information are key to the survival of every bird and can also alert the animal to danger associated with invasion of their body.

Somatosensory receptors almost certainly provide information about objects taken into the mouth, and the presence of these receptors throughout the body suggests that they are also vital for monitoring internal body conditions. The bill-tip organs found in certain bird species show, however, that the somatosensory information can also guide highly specialised behaviour in foraging. Crucially, bill-tip organs can make the bill a semi-autonomous tool for actively exploring the environment, not just a device that is guided by vision for seizing and manipulating objects. Furthermore, the bill-tip organs of parrots also provide a vital source of specialised information that underpins the use of their bills as a 'third limb', guiding these birds in their exceptional ability to climb, both up and down, in structurally complex situations.

Taste

Taste is big business. Every year around the globe thousands of books are published, TV programmes are broadcast, thoughts are posted on social media, all based around taste. There is a sophisticated, though often confusing, language used to describe the tastes of foods and drinks. Experts claim to taste the presence of all sorts of things in the 'notes' of wines. Some people carve out careers as tasters, and some are labelled 'super-tasters'.

The problem for the majority of people is that all the jargon about taste is a mystery that confuses rather than clarifies. We may get the impression that some people are trying to hoodwink us. As with other senses, there is a fundamental problem that stems from the fact that we generally perceive a unified sensation when we taste something, and do not naturally break it down into constituents. However, as with touch and olfaction, taste arises from the responses of populations of different receptors which detect separate properties of a stimulus.

Taste is similar to olfaction in that it is a sampling or screening process that detects the presence of specific groups of chemical compounds. The receptors of taste are of relatively few types, but they can be numerous and occur in different concentrations in different locations within the body. It is the combined responses of these different receptors which give rise to the overall perception of the taste of something. As with olfaction and touch, there are marked differences between species in the absolute and relative numbers of different receptors, and in their distributions within the body. This differential distribution of taste receptors strongly suggests that natural selection has shaped both which tastes are more important for different species and how they are sampled in different species. Thus, there is a sensory ecology of taste.

While we may be sceptical of the hype about the taste of different foods and drinks, it is clear that taste is vitally important. Taste in animals, including birds, has evolved primarily to do two things: to identify nutritious foods and to detect potentially poisonous substances. These two aims are achieved through the rapid detection of key nutritionally relevant compounds such as carbohydrates, amino acids, lipids, salts, and calcium, and the equally rapid recognition of toxic compounds.

The distribution of taste receptors in the mouths of birds follows the ingestion pathway, and this allows items taken into the mouth to be rapidly screened for a range of chemical properties as they are transported through the mouth cavity, before entering the gastrointestinal tract or gut. As a pet owner you may be disappointed that your animal does not savour the taste of a treat, but just gulps it down. But subtle aesthetics are not what the sense of taste is for; it is for the rapid screening of items as either potentially harmful or nutritious. Once that screening test has been passed the food should be ingested rapidly, in case there is more to be had that a rival might grab.

Much of what is known about taste in birds comes from studies of domesticated breeds, notably chickens, ducks, and turkeys. This is because understanding taste in these birds is valuable for determining how to maximise the efficiency of food intake in commercially exploited species. Outside of these domesticated species, some taste research in birds has also been driven by the search for tastes that birds find aversive, with a view to developing bird repellents.

Taste buds

In all vertebrates, taste signals are generated in taste buds. These contain groups of receptors that detect the presence of specific chemical compounds. Like other sensory receptors, taste receptors have a very long evolutionary history and show basic uniformity of structure and function across diverse animal groups. Taste buds are groups of receptor cells held within flask-shaped pits embedded mainly within the skin of the tongue but also in other structures within the mouth. They are also found at sites well down through the gut. These are probably involved in many aspects of sensing foods as they are digested and may be particularly important in the screening of foods after ingestion. They are probably responsible for inducing vomiting following the ingestion of a noxious substance.

Taste buds link to foods and liquids through a pore. The individual receptors of the taste bud sample the chemical composition at the pore entrance. The types of taste buds in birds are not clearly established, but at least three are found, classified primarily by the shape of the pore and the types of receptor cells which extend to the pore entrance.

The first clear description of taste buds in birds was relatively late compared with other senses. This may have been because a very influential anatomical text by the German anatomist Friedrich Sigmund Merkel had stated in 1880 that taste buds did not occur in birds; that was sufficient to put off other researchers looking for them. However, in the first decade of the twentieth century careful descriptions of taste buds in Mallards *Anas platyrhynchos*, Common Starlings *Sturnus vulgaris*, and parrots were published by Wolfgang Bath and Eugen Botezat, although they disputed whether taste buds occurred on the tongue. However, it seems that this was because they differed in their definition of what constituted the tongue in different species. It has subsequently been shown that taste buds in Mallards occur not only on the tongue, but also near the bill tip (just behind the bill-tip organ) and in the floor of the mouth just behind the bill tip.

This positioning of taste buds shows that in Mallards, and probably many other bird species, the taste of objects held at the bill tip can be sampled. This is important because it indicates that an object does not have to be brought far into the mouth for its chemical composition to be sampled and hence screened for its value as a food item. In all birds so far investigated, taste receptors occur on the tongue, particularly towards the back, but they do not occur widely over the upper surface of the tongue, as is typical in mammals.

Taste receptors and taste genes

Although the early understanding of taste receptors in birds relied on careful anatomical work and behavioural experiments, those kinds of investigation are no longer essential to gain an idea of which types of taste receptors are present in a species. This is due to the identification of 'taste genes' in a wide range of vertebrates. Taste genes are sections of genetic code that indicate that certain types of receptors are present in the animal, which in turn provides good evidence of the ability to detect certain chemicals in a particular species.

Taste genes seem to be highly conserved across vertebrate classes; that is, they show little variation across diverse animal groups, attesting to the critical role that taste has played in the survival and adaptation of species throughout their evolutionary history. This work has also led to the identification of taste receptors, as opposed to the taste buds in which the receptors are housed. This genetic approach has shown that in mammals taste receptors occur in many locations in the gut and also in structures involved in the absorption of nutrients including liver, adipose tissue, and hypothalamus. There is some evidence that taste receptors occur in similar locations in birds. As with somatic sensitivity, once key receptor types had evolved, they have been retained over time with little elaboration. What have changed are the relative numbers and distributions of receptors within the bodies of different species.

How acute is taste in birds?

Definitive investigations to quantify the sensitivity to taste are difficult to carry out, and clear answers about threshold concentrations for the detection of specific tastes are usually not available. What has been clearly demonstrated is that birds can detect a certain taste at a range of concentrations and that they may show both aversion and preference depending on concentration, although the threshold for detection is usually not known. However, it does seem that taste in birds is much more sensitive than was once assumed, and that taste sensitivity to different compounds is widespread among birds.

Mainly because birds do not chew, it was thought that chemicals from items in a bird's mouth were not readily exposed to taste sampling. It was also thought that taste buds occurred in relatively low numbers. However, when the number of taste buds is considered in relation to the size of food that is taken into the mouth in one bite, and to the volume of the mouth cavity, it seems that the number of taste receptors in birds is no lower than in mammals. This suggests that the thresholds for taste detection in foods could be similar in birds and mammals.

The tastes of birds

It seems likely that birds have at least seven distinctive tastes, although older research refers to just four main tastes: sweet, salt, sour, and bitter. Of the seven tastes currently considered to be present in birds, five seem to function primarily in driving particular appetites for specific nutrients in the diet, while the remaining

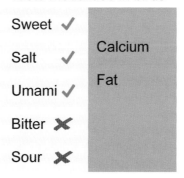

FIGURE 7.5 Schema of the primary taste modalities of birds. Those in the yellow box are well established and are divided between 'appetite tastes' and 'warning tastes'. Appetite tastes drive the ingestion of food items beneficial for nutrition, while warning tastes serve primarily to advise against the ingestion of items, perhaps because they are contaminated with toxins. Those in the pink box may drive specific appetites, but evidence for them is more circumstantial and specific receptors have not been identified.

two function primarily to provide warning signals about food items that are likely to be detrimental to health (Figure 7.5). When behavioural evidence on the tastes of birds is placed alongside evidence of taste genes, it seems safe to conclude that birds are well able to discriminate the same range of tastes as mammals, including ourselves.

Sweet

Faced with a choice between liquids containing sugars or not, omnivorous and granivorous birds (chickens) do not respond positively to sugars; that is, they may reject further offerings of sweetened liquids or they may seem oblivious to the presence of sugars. On the other hand, nectivorous (hummingbirds) and frugivorous bird species (e.g. Cockatiels *Nymphicus hollandicus*) respond positively to sugars. Taken together, these results suggest that birds do taste sweetness, and that there is a clear sensory ecology for this sweet taste – i.e. the response to the presence of sugar molecules in food varies with the diet and ecology of the species. Common Starlings show a preference for fructose and glucose, and this may be related to their preference for fruits that are ripe, and we might therefore expect a wide range of other birds which eat fruits to also show such preferences.

Umami or amino acids

While humans find it relatively easy to judge whether something tastes sweet, the umami taste seems to be something that we may need a little training to identify. However, this may be a cultural or learning phenomenon. It may depend on early exposure to sweet substances and encouragement to value sweet foods as treats while food items that taste of umami are not usually the stuff of treats.

Umami is often identified as the taste of monosodium glutamate or glutamic acid, found at high concentrations added to many prepared foods that would otherwise taste rather bland. However, umami is a taste whose function is the detection of protein or amino acids. Detecting these substances is just as important and fundamental to good nutrition as sweet taste is for the detection of sugars and assessing caloric value.

Umami has been shown to play a key role in guiding the choice of foods in birds. For example, in choice experiments Chicken *Gallus gallus domesticus* chicks showed a preference for a diet containing amino acids. The actual substances detected by umami or amino acid receptors in birds require further investigation. One candidate is the chemical compound L-alanine. It is detected by both Common Starlings and Red-winged Blackbirds *Agelaius phoeniceus*. It is suggested that umami receptors in Chickens may be capable of detecting a wide range of amino acids and that they are driving a protein-specific appetite. It is also thought that umami receptors are involved not just in the detection of food quality but also in the coordination of post-ingestive and metabolic events in the gut; this is because umami receptors have been identified in the hypothalamus, liver and abdominal fat tissues of Chickens.

Salt

The taste of salt (sodium chloride, NaCl) is one that humans readily distinguish, and which seems capable of driving a specific appetite, leading some people to seek out foods with high salt content, while at other times salt may be avoided. This seems also to be the case in birds, and two quite different behavioural responses have been recorded depending upon the concentration of salt presented. Most is known about this in Chickens because of the need to provide the correct diet in birds raised for human food. High concentrations of salt elicit aversion, but low concentrations usually produce an attraction response, particularly after birds have been fed a diet deplete in sodium. This supports the idea of a specific appetite for salt and is also evidence of a fine discrimination of salt concentration. Furthermore, it seems that this is continuously calibrated against the internal state of the bird, something which probably also happens with appetites driven by other tastes such as umami and calcium.

As with mammals, high concentrations of salt (>2% solution) are toxic. However, a number of birds, most notably those that spend long periods at sea, have specialised salt glands situated above their eyes that allow the birds to ingest sea water but excrete the salt at high concentration through their nostrils. The salt glands provide an additional mechanism to the kidneys for extracting salt from the body.

Sour

Sourness is related to the acidity of food or liquids. The presence of acids is often caused by bacterial fermentation, and so detecting a sour taste can act as a powerful screen for foods that are not safe to ingest. In mammals the taste of acid at higher concentrations can evoke a strong rejection response. However, as with salt, a graded response to acid/sour is shown depending on the degree of acidity. Chickens are tolerant of medium acidic or alkaline solutions (either side of the neutral pH value of 7), but they avoid extreme pH values.

Although the taste-sensitive cells for both salt and sour are known in mammals, they have yet to be identified in birds. However, genes for these receptors have been shown to be present in the Chicken genome, suggesting that these two tastes are similarly detected in birds and mammals.

Bitter

Detecting bitter seems to be crucial for everyday survival because it indicates what not to eat. This is because bitter tastes are usually associated with naturally occurring toxic compounds of plant or animal origin, especially insects. Bitter-tasting substances are widespread and include some relatively common compounds such as tannins and phenylpropanoids, but also many more. These seem to be aversive to birds. For example, Common Starlings and Red-winged Blackbirds may reject substances containing such compounds on first exposure, or they will reject them on a second exposure if they resulted in the birds feeling unwell after the first ingestion – a good example of one-trial avoidance learning. There have been attempts to exploit this aversive response to bitter in the development of bird repellents. Quinine is a compound known particularly for its bitter taste to humans, and Chickens and Cockatiels also show a strongly aversive response to quinine. It has been shown to be sufficiently unpleasant to Chickens that they learn to avoid it after just a few exposures.

Calcium

A specific appetite for calcium seems to occur in birds, particularly during egg-laying. At this time calcium is required for the construction of eggshells. A specific appetite, or at least a dietary preference for calcium-rich items such as shells, bones, and calcareous grit has been well documented in many birds. Chickens are among birds which exhibit a well-defined appetite for calcium during their early development when the skeleton is growing rapidly. A similar appetite also occurs in adult female Chickens, especially those used in commercial egg production, presumably because of the high demands for calcium for the formation of eggs over extended periods. While there does seem to be a specific appetite for calcium, it is not clear how important taste is in driving this. Recent work has suggested that 'calcium-like' sensing is associated with a distinct taste, but conclusive evidence is required to be sure of this.

Fat

Recent evidence suggests that in mammals a fat-sensing taste receptor exists, and that this plays an important role in regulating the caloric content of the diet. Certainly, there is evidence that people deprived of fat can develop strong appetites for it. However, whether it depends on a specific receptor is not clear, and to date no apparent fat taste receptor genes have been found in bird genome databases. It is possible that birds do have a receptor for fat and this may be in the gut rather than in the mouth.

Conclusion: taste in birds

Like other senses, taste is based upon a relatively small number of different types of receptors. And again, as with other senses, there can be marked differences between species in the relative numbers of the different receptors and where they are positioned within the body. This means that there is flexibility in the ways that different species gain taste information and in the uses that they might put it to.

Fundamentally, the taste system of birds can be viewed as a group of nutrient sensors which have evolved to evaluate the quality and nutritional content of foods, primarily at the point of ingestion, but also at various points post-ingestion. Birds are using taste continually, since they must assess every item of food that they ingest. There is some evidence that the taste systems of birds exhibit distinct relationships with the ecology of the species and that the exact array of taste receptors in any one species may be closely linked to its diet and the means of ingestion. However, much more comparative work is required to develop a full understanding of the sensory ecology of taste in birds.

Taste seems to have a role that is highly influential in the control of the foraging of birds. For example, the pathway of a food item through the mouth of a Mallard is highly linked to the taste receptors that screen the food for its qualities. There is also evidence that some birds use taste in a more sophisticated way that makes their foraging more efficient by directing them to places where prey may be profitably detected. This evidence comes from shorebirds of the genus *Calidris*. They have been shown to use taste cues to determine concentrations of their preferred prey buried in mud and sand.

Calidris species are among those that forage by probing in soft substrates, using tactile cues to locate individual items. However, when faced with an expanse of apparently uniform mud or sand, how do the birds know where to focus their probing efforts? A series of intriguing experiments involving four species – Sanderling *C. alba*, Dunlin *C. alpina*, Purple Sandpiper *C. maritima*, and Red Knot *C. canutus* – showed that these birds could distinguish between patches of sand which contained 'taste' and 'no-taste'. Taste came from secretions left by polychaete worms or bivalve molluscs which were introduced to washed sand but removed before the experiments. No-taste was simply the washed sand. It was found that all birds spent longer and probed more frequently at the patches containing 'taste' than at patches without taste. It was concluded that in all species taste sensitivity is a cue that guides foraging, but that its importance probably varies between species, being of most importance in Dunlin and least in Red Knot.

This example of using chemical information from taste receptors to find locations in which foraging may be more profitable parallels the examples discussed previously in which seabirds and passerines are able to determine profitable foraging locations, but not individual food items, by using olfactory cues (Chapter 6). In the case of olfaction, cues are gained at a distance. In the probing shorebirds cues are located at close proximity, but they are achieving the same end: more efficient foraging.

Chapter 8

Sensing the earth's magnetic field

It is difficult to find the appropriate place in this book to discuss magnetoreception. It is neither telereceptive nor intimate. Magnetoreception sits on its own; unlike the other senses it does not ultimately provide information about objects. Rather it seems to provide information that is unique to each individual: its current location and the direction towards another location.

Magnetoreception provides this information by detecting properties of the earth's magnetic field, the geomagnetic field, which varies continuously across the globe. The geomagnetic field pervades a bird's entire body, in fact it pervades the bodies of all organisms. It pervades our own bodies, but to date there is no convincing evidence that we can detect it, although there have been claims that some people can detect the earth's magnetic field and use it for direction-finding.

There is evidence, however, that some birds, and a wide range of other animals and some bacteria, do detect the earth's magnetic field. There is even the suggestion that detecting the geomagnetic field is an ability common to all birds, regardless of whether they are long-distance migrants or sedentary species that move only within a limited home range. Magnetoreception may be the key source of information that enables birds to make predictable movements towards target locations at large spatial scales, sometimes between locations that are hundreds of kilometres apart, but also at much smaller scales, for example around a territory or within a home range.

To humans, magnetoreception is mysterious. Mysterious, because we have no direct access to it. We can only try to imagine what it is like to sense a magnetic field; we have nothing within our experience to compare it to. Metaphors based upon other senses would seem to be pointless. After all, knowing what it is like to feel something can give us no idea what it is like to taste it. The information and the experience of each sense is unique.

Magnetoreception is also a mysterious sense because although its presence has been deduced in birds from behavioural experiments, the receptor, or receptors, has yet to be identified. For all other senses we are able to look at diagrams of the receptors and associated sense organs, and we can get a grasp of their form and function. But this is not yet possible with magnetoreception. There is some evidence of mechanisms that might be capable of detecting the earth's magnetic field but no definitive demonstrations of where they are situated or how they might work. This is both frustrating and intriguing.

Magnetoreception is relatively new to science. The first definitive experiments indicating that magnetic fields can influence the behaviour of an organism, a bacterium, were not published until 1963, and work to systematically investigate the magnetic sense began only in the 1970s. This first demonstration in bacteria was in fact a demonstration of magnetotaxis: the bacteria were pulled in a specific direction by a magnetic field, but they were not sensing it. They were passively responding to its presence, dragged in one direction by an applied magnetic field.

Since the 1970s a large body of work has demonstrated or deduced the importance of information about the earth's magnetic field in many animal taxa, including birds. The first work on birds was done in homing pigeons. In these studies, it was possible to show that the ability to home could be disrupted if the geomagnetic field around a bird was disrupted. This was achieved by fixing either a magnet or an electrical coil to a bird's head prior to release on a homing flight. The birds had first got used to making homing flights wearing similar devices that did not disrupt the magnetic field. However, disruption of homing ability did not occur under all sky conditions, suggesting that the homing of pigeons does not rely exclusively upon magnetic field detection. Early researchers had hoped that magnetic field detection would be the key to pigeon homing, but such a simple conclusion could not be made.

Detecting the geomagnetic field and extracting information from it can be thought of as analogous in function to the now ubiquitous GPS (Global Positioning System). It seems that the geomagnetic field can be used by a bird to provide information about both its current position and the direction towards a particular goal. There is a body of evidence that birds do use the geomagnetic field to determine at least their initial orientation when departing upon migratory flights, and also by homing pigeons when setting out to return to their lofts.

Although important, magnetoreception is, however, just one cue that may be used by homing and migrating birds. There are, for example, compasses used by birds that rely on the stars and the sun. Homing pigeons also seem to employ a number of different cues based on visual landmarks and olfaction (see Chapter 6), as well as possible compasses. The cues that are used in particular instances depend on weather, location, and the specific experience of individual birds.

Taken together, all of this evidence suggests that detecting the earth's magnetic field is just one component that birds may use to achieve the complex tasks associated with homing and navigation. The earth's magnetic field may not be the most important source of information for the conduct of these tasks as was once assumed and perhaps hoped. People once wrote about a somewhat magical 'sixth sense' to explain the long-distance movements of migrating and homing birds. When magnetic field detection was first described, it was hoped that it would provide a simple explanation for a complex set of behaviours.

Direction towards a goal or away from a particular starting point can be derived from the horizontal component of the earth's magnetic field (the geomagnetic vector), and this is the information that humans have learned to detect through the invention of the technological compass. Location on the globe can theoretically be derived from the total intensity and/or inclination of the geomagnetic field vector. This aspect of the earth's magnetic field exhibits gradients between the earth's magnetic poles and the magnetic equator. Although humans have technology to determine this, it has not been commonly employed in human navigation devices.

Species sensitive to the earth's magnetic field

Demonstrating sensitivity to the earth's magnetic field has come from evidence that an animal is able to detect one or both types of geomagnetic information, direction and location. Animals which seem to have access to a magnetic compass include species of the phyla Mollusca and Arthropoda, plus species in all major vertebrate groups including cartilaginous and bony fish, amphibians, reptiles, mammals, and birds. Magnetic field information is known to be used for the orientation of colonies by bees and termites. It is also thought to provide directional information in the extended migrations of eels, salmons, marine turtles, and birds. In birds, magnetic cues are known to be used to determine direction at different spatial scales, including within a home range.

Among birds, magnetoreception could be a widespread ability, since it has been demonstrated in birds from three avian orders including homing pigeons (Columbiformes), domestic chickens (Galliformes), and a number of species of passerines. However, the actual number of bird species in which it has been demonstrated is small and certainly not comprehensive (Figure 8.1).

Compass mechanisms

Based on the earth's magnetic field, two types of magnetic compass mechanisms are possible. The first type is a *polarity compass*, which works in a similar way to a technological compass, using polarity of the magnetic field to distinguish between magnetic north and south. The second type is an *inclination compass*, and this appears to be the type of compass used by birds. It relies on the axial course of the geomagnetic field lines, with directional information obtained by interpreting the inclination of the field lines with respect to up and down (derived from gravity). This mechanism distinguishes between 'pole-wards', where the field lines run downwards, and 'equator-wards', where they run upwards.

FIGURE 8.1 Birds species which have shown positive results in investigations designed to tease out how birds detect the earth's magnetic field. Rocks Doves *Columba livia* and Bobolinks *Dolichonyx oryzivorus* (top) have shown evidence for detection based upon the 'magnetite model', possibly involving deposits of magnetite around the olfactory nerve. European Robins *Erithacus rubecula*, Silvereyes *Zosterops lateralis*, and Garden Warblers *Sylvia borin* (bottom) have all shown evidence for detection based on the 'radical pair model', which suggests that there is a mechanism based in the retina, but in the right eye only. (Photo of Dove by Muhammad Mahdi Karim, www.micro2macro.net [GNU free documentation licence], Bobolink by JanetandPhil [public domain], Robin by Sergey Yeliseev, Flickr, Silvereye by Noodle snacks, own work [CC BY-SA 3.0], Warbler by Billyboy [CC BY 2.0].)

Detecting the geomagnetic field

The way in which birds detect the earth's magnetic field is unclear. What is, or are, the magnetoreceptors of birds is still debated. However, a consensus now seems to be established that there are likely to be at least two detector mechanisms, with all types of mechanisms probably occurring in the same individual. It is possible that one mechanism is more concerned with determining location while the other is concerned with direction, but this is by no means certain.

The two main mechanisms are referred to as the *magnetite model* and the *radical pair model*. The use of the word 'model' is important, since it indicates that these are ideas of what might be, rather than describing actual mechanisms and their receptors. A third mechanism positioned in the inner ear, in a structure called the lagena, has also been proposed. But the basis of this mechanism's ability to detect the magnetic field is not known.

The magnetite model proposes a primary process involving tiny crystals of permanently magnetic material located in the head. The radical pair model proposes a so-called 'chemical compass'. It is based on a complex chemical process called *singlet–triplet transitions* that occurs in particular photopigments in the retina of the eye when exposed to light. However, this is a process quite distinct from vision.

The magnetite model

Magnetite crystals were first found to function in the magnetotaxis of particular bacteria (magnetotactic bacteria), and similar crystals were subsequently reported in birds. In Bobolinks (Icteridae) deposits of iron oxide (probably magnetite) have been described (Figure 8.1). These lie in sheaths of tissues around the olfactory nerve, the olfactory bulbs, and between the eyes, and also in bristles which project into the nasal cavity. Very small magnetite crystals have also been found in the skin of the upper bill in Rock Doves. There is evidence that the nerve fibres which run from these regions where magnetite has been identified are responsive to changes in earth-strength magnetic fields in Bobolinks. This anatomical and physiological evidence has been supported by various behavioural experiments in which birds were placed in magnetic fields which overwhelmed the low-strength geomagnetic field. This treatment disrupted the orientation behaviour of birds, which was reinstated when the high-strength magnetic field was removed.

The radical pair model

Evidence in support of the radical pair model in the eye was first found through demonstrations that magnetoreception in homing pigeons and in migratory birds is light-dependent. Species in which this has been demonstrated are passerines which have migratory populations; these species are from three families, European Robins (Muscicapidae), Silvereyes (Zosteropidae), and Garden Warblers (Sylviidae) (Figure 8.1). The radical pair model proposes that magnetic fields are detected by specialised photopigments in the retina, with the most likely candidates being cryptochromes. These are flavoproteins that are sensitive to blue light and are found widely in plants and animals, where their primary function is in the control of circadian rhythms. Although they occur in the retina they are not located within the rod or cone photoreceptors.

Magnetoreception in birds is not only light-dependent but depends also on the wavelength of light and on the eye that is used. Evidence to date suggests that radical-pair-based magnetoreception is found only in the right eyes of birds. Furthermore, experiments with Rock Doves and Chickens have shown the presence of a magnetic compass that works only under blue-green light (up to a wavelength of 565 nm), but normal migratory orientations are disrupted under yellow light (light with wavelengths longer than 582 nm). This link to blue light is another indication of the involvement of cryptochromes.

Conclusion: magnetoreception in birds

The above picture of magnetoreception is complex and has generated some controversy in recent years. A consensus is yet to emerge on how birds might use magnetoreception and what its detecting mechanisms are. Researchers have been known to disagree quite vehemently about possible mechanisms. This may in part be because humans have difficulty in comprehending what the perception of magnetic fields could be like. We have no personal reference for what is being discussed. There is extra confusion perhaps because it seems that one aspect of magnetoreception in birds may occur in the eye, and there has been a temptation to imagine that birds must in some way 'see' the geomagnetic field. However, there is no reason that this should be the case. That another aspect of magnetoreception may occur in or around the olfactory nerve does not mean that bird might 'smell' the geomagnetic field.

There is a good body of evidence that birds, along with many other animals, are able to extract information from the earth's magnetic field. It is well established that birds are able to employ such information to determine migratory orientation at the time of departure, and probably use this within a suite of vision-based information (landmarks, star- and sun-based compasses) and olfactory information to guide homing.

It also seems that magnetoreception can play a part in the control of bird movements at smaller scales, perhaps helping them to find their way back to sites within a home range or territory. As such, magnetoreception is a sense that may be employed in birds continually, not just referred to at key times, for example when birds home or migrate. Like other senses, magnetoreception is likely to be permanently active, not switched on and off for the execution of particular tasks.

While many fundamentals remain to be determined, it does seem clear that more than one type of magnetic information is used by birds, and that two, and possibly three, types of detector mechanisms may be involved. Perhaps birds use cryptochrome photopigments in the right eye for recording magnetic directions, and magnetite-based receptors in the upper bill for recording differences in magnetic intensity, and also use a mechanism in the inner ear, although its basis is not known.

One pair of prominent researchers in the field of bird geomagnetism, Wolfgang and Roswitha Wiltschko, famously proposed that birds seem to have a compass in their right eye and a magnetometer in their bill. This is an attractive idea, but it is by no means proven. Magnetoreception in birds is still a mystery. There seems no doubt that it exists, but exactly what it is used for and the mechanisms that extract information from the environment are still uncertain. Too few birds have been investigated for it to be possible to establish a 'sensory ecology of magnetoreception', in which magnetic field information can be described as tuned to the challenges of different environments for the execution of different tasks, in different species.

Chapter 9

Birds in the dark

For most modern humans, darkness and night-time present major problems. In fiction, night-time is the setting for all kinds of misdemeanours. From the appearance of a ghost to the conduct of a foul murder, night-time is the clichéd setting. Things are always going bump in the night, and we learn from an early age to be wary of them.

For modern humans, the easiest way to deal with these apparent problems of night-time and darkness is simply to banish them. Switch on the room lights, switch on the streetlights, switch on the car headlights, these are the instant solutions offered to us. Satellite images of the world at night show that switching lights on is a universal solution to night-time, so much so that this solution has itself led to a new problem, light pollution. It is pollution that stops us seeing the Milky Way, and it can disrupt the natural behaviours of birds and much other wildlife. Where do our modern human problems with darkness and night-time come from? Do other animals have the same problems with night-time? Do we simply not understand night-time?

At root, the modern problem of night-time for humans arises because we do not wish to acknowledge the constraints of our own sensory ecology. We view ourselves as daytime animals, proud of what we can do during the day, and we wish to be daytime animals until the very moment that we choose to go to sleep. We cannot escape the requirement of sleep, but we want to be daytime animals until our heads hit the pillows.

We hate to be 'in the dark' about anything. Being in the dark is a metaphor that springs from our sensory ecology. It acknowledges that after the sun has gone down the information that we can retrieve from our environment seems to be reduced, so much so that we may no longer feel in full control of our situation. The problems of being in the dark, however, also stem from the fact that most, if not all, modern humans no longer have experience of natural night-time. Most people are unaware of what they might actually be able to do at night. For most people, night-time is simply defined as the time when you have to put on a light; it is not something to be understood or experienced. Most people let the cultural myths and assumptions about night-time become their knowledge of it. Venturing outdoors at night without an artificial light is now regarded as definitely problematic, probably eccentric, and possibly suspect.

I was lucky and got my first taste of natural night-time at an early age. For me night-time became a fascinating time – and this early experience of natural night in a woodland probably laid the foundation for my lifelong interest in senses and sensory ecology.

I was 10 years old when I was taken on my first night ramble. I had lain in bed and heard owls hooting in a local wood. During the daytime the wood was quite a magical place, and my clearest memory of it was that it was full of oak trees and bluebells. It was later designated as a Site of Special Scientific Interest, but to us kids it was just a local spot to go and play. I had asked my father about the hoots and screeches that I could hear from my bed. To my surprise, and concern, he suggested that we must go and find out. A scary prospect, especially as he said I was not allowed to take a torch. I was also told that I would have to be very quiet.

I am not sure what I expected, but by the time we had walked the 10 minutes from home to get under the tree canopy, and close to the hooting and screeching, I was amazed. My eyes had begun to dark adapt, but I could not see everything. Tree trunks were just large shapes, their leaves and branches outlined indistinctly against the sky, and of course there were no colours. However, as we stood very still just looking and listening, two things became apparent. The longer I stood there, the more I could see, but the woodland that I experienced was not simply the woodland of daytime without light. It was a different place, with lots of subtle noises. Leaf litter rustled all around, and the more I listened the more I heard, and the more I looked the more I could see.

I was tuning in to the sights and sounds of a night-time woodland. We did not see the owl or a fox, but we heard them nearby. I was hooked on this very different experience. It became something special to me, shared with my dad, and we visited together a number of times. I realised that this was the place where a lot of animals lived out their lives, quite different from the animals that we might glimpse or hear during the day. Clearly this was home to other animals, and I too felt at home there once I had learned to relax and tune in. But in fact, tuning in was not difficult. It was just a matter of time, not looking at bright lights, but staring into the dark, allowing my eyes to fully dark adapt for at least 30 minutes.

Ten years had passed when the lucky opportunity to start research into the senses of owls came my way. I had no doubt that this was the PhD project for me. I was launched on a career of looking into bird senses. Ever since those first experiences of natural night-time I have enjoyed just experiencing that world. Even though I have measured it, and measured many aspects of bird senses, especially in relation to low light levels, the experience of simply being out in the countryside at night retains a magic. There is always something to hear, always something to see, something that is not available during the day. Many people might suggest that we only see things properly during the day, but that begs the question of what is proper. Whether a scene is brightly coloured and contains fine detail, or is monochromatic and composed of larger indistinct shapes, both present vital aspects of the world.

If the flight behaviour of birds depends upon detailed visual information, it would seem obvious that the large majority of birds should go to roost as soon as light levels start to fall when the sun dips below the horizon. But many birds only begin their daily rounds of activities at dusk, and many species complete specific parts of their annual cycles, especially migratory flights, overnight. Some birds forage at very low light levels but live the rest of their lives at higher daytime light levels. Do these birds have special sensory tricks?

We need more than wisdom to explain the activities of these birds at low light levels. The myths of great wisdom have often chosen nocturnal owls as their symbols. The Greek goddess of wisdom Athene, and her Roman successor Minerva, were both symbolised by an owl. Owls turn up in similar symbolic guises in many other cultures, including those of ancient Egypt and the Americas. Even today, owls are treated popularly with an element of awe or reverence. The voice-over becomes earnest and measured when an owl appears in a natural history film. An owl never plays the comic role. They are always wise and measured in children's literature, perhaps best captured in the verse:

A wise old owl sat in an oak
The more he heard the less he spoke
The less he spoke the more he heard
Why can't we all be like that wise old bird?

While the key message of the rhyme is that wisdom is gained mainly through quiet and concentrated thought, rather than activity, it also hints at a trade-off in information between two different senses, vision and hearing. This is a key idea which will be touched upon a number of times in this chapter.

Is night-time a problem?

When I first experienced natural night-time all those years ago I was learning about some of the most fundamental properties of vision. Of course, the learning was experiential, and it was many years before I got to grips with the scientific explanations behind what I had experienced. In Chapter 4 we saw that as light levels decrease, spatial resolution and contrast sensitivity inevitably fall (Box 2.3 and Figure 4.8). Furthermore, colour vision does not function at lower light levels. These changes in vision are caused by fundamental constraints that apply to the vision of all vertebrates. They arise because of the physical nature of light itself, and because of fundamental limitations on the physiology of the image analysis system of an eye, the retina.

These decreases in spatial resolution with light levels mean that there is a gradual reduction in the information that can be gained through vision about objects and surfaces that surround an animal. As light levels decrease only larger objects such as the trunks and branches of trees can be detected with certainty. Also, objects that contrast highly with their backgrounds, especially the sky, become more significant. Spatial detail, even close at hand, may be lost as light levels of a woodland floor can fall below the absolute threshold of vision (Figure 9.1; see also Box 9.1).

There are many implications for the guidance of nocturnal activity in these facts. There is certainly less information available compared with what will be available during the day at the same location. The decrease in resolution can also be considered to have an important consequence for the distances at which objects of particular size or contrast can be detected. For example, if an object is just detectable at a particular distance at a high light level, then when acuity decreases by 10-fold the same object will have to be 10 times closer for it to be detected.

It is probably this that we consider the most problematic aspect of continuing our daytime activities at night. It is the reason why we reach for the light switch, the reason why we feel that being in the dark is a problem, because it means that

FIGURE 9.1 The transition from day to night simulated in three photographs of a bluebell wood. In daylight this is the hunting domain of Eurasian Sparrowhawks *Accipiter nisus*, but as light levels fall these birds cease activity and Tawny Owls *Strix aluco* start to hunt. The changed appearance of this scene as light levels fall, especially the loss of spatial detail and colour, is due only to properties of vision. The physical characteristics of the woodland have not changed, only the amount of ambient light has decreased. Even the spectral distribution of ambient light is the same under daylight and moonlight. Moonlight is simply sunlight reflected from the moon's surface. Sparrowhawks and Tawny Owls exploit different resources within the same habitat, and to do so employ different foraging strategies underpinned by different information.

we cannot detect at the right distance those things which seem essential when we act during daytime. Basically, the nocturnal world becomes more intimate, the world presses in. However, would this be a problem if the distance at which we could detect things was always relatively close, would we then see night-time as a problem? Presumably not, because our behaviour and the tasks that we expect to do would simply be different. What this suggests is that night-time is a problem for a diurnally active animal, but is not necessarily a problem for a nocturnally active one. Nocturnally active animals just have different tasks and use different sets of information for their conduct. They live in a different world and so use different information to guide different tasks.

So, in what ways are owls wise? It seems that the 'wisdom' of owls has three aspects. First, it lies in relying on information about specific kinds of objects, for example detecting tree trunks and branches, rather than leaves. Second, the wisdom to use senses for the control of a more restricted suite of behaviours compared with what might be expected in birds that are active in daytime. Third, the wisdom to rely on information from more than one sense for the guidance of key tasks. All of these aspects of the activities of birds at low light levels will now be explored. It will be seen that these wisdoms are not exclusive to owls.

Is night-time a problem? It is only a problem if viewed through the perspective of a daytime active animal. Clearly, for nocturnal animals their activity presents no more challenges and problems than are faced by diurnal animals, the challenges are just different. It is possible for humans to embrace the wisdoms of owls. However, to do so we need to abandon the idea that what we do during the day is the exclusively important stuff of our lives. To be active in natural night-time, guided by what information our senses can provide, is part of human experience. It is experience that is no less rich than that obtained on the brightest and sunniest days.

Getting to grips with night-time

Night-time is complicated, certainly more complicated than daytime. Night-time starts when the sun goes below the horizon. As the sun falls further below the horizon ambient light levels change rapidly through various levels of twilight. Light levels then plateau through the night before they start to rise through the morning twilight and the sun eventually reappears above the horizon. However, the light levels at which nights plateau will vary depending upon latitude and time of year, and these two factors also affect the length of the night. Light levels will also vary from night to night due to the presence and phase of the moon, the presence of clouds, and, beneath a woodland canopy, the extent of leaf cover. This all adds up to a light regime that is much more variable than during daytime, and more difficult to predict from night to night and from place to place.

In essence, daytime is a relatively stable light environment; night-time is not. Furthermore, the magnitude of variations in light levels at night is huge. The total difference between the light available under clear skies on the brightest days and on the darkest nights at mid-latitudes is just over 100 million-fold ($\times 10^8$) (see Box 2.4).

The important effect of light-level variability during the night at any one location is that the spatial resolution of an animal's vision must be changing continuously by non-trivial amounts. Consequently, the spatial information, and hence the detectability of objects and surfaces, also changes continuously. Furthermore, as Box 2.3 shows, the way in which spatial resolution changes with light level is not linear. Daytime light levels can change over a thousand-fold range (Box 2.4), yet spatial resolution changes relatively little. However, over the million-fold range that light levels can vary at night, resolution changes are much more marked. These changes in acuity with light level have the important consequence that an animal which is exclusively active during daytime will experience relative stability in its spatial resolution, and hence relative stability in its potential to gather visual information from its environment. On the other hand, a nocturnally active animal at the same location will experience much greater, and less predictable, changes in its ability to retrieve spatial information from night to night.

Absolute visual sensitivity and the challenges of the nocturnal environment

High visual sensitivity is the obvious solution that explains how animals can be active at low light levels. Put simply, absolute visual sensitivity in nocturnally active species should be higher than in diurnal species. In nocturnal animals it would be predicted that natural selection has driven both the optical structure of the eye, and the physiology of the retina, to trap as much light as possible. This does in fact seem to be the case. There are, however, some surprises when the eyes of diurnal and nocturnal birds are compared, and also when they are compared with our own visual sensitivity. High sensitivity alone cannot explain nocturnal activity.

The ways in which visual sensitivity can vary between eyes were discussed in Chapter 4. From that discussion it can be predicted that eyes which have evolved to maximise light capture, and hence sensitivity, should be absolutely large. The focal length of the eye should not increase in proportion to the overall eye size, however. This is because the important determinant of image brightness is the f-number. This is a familiar term to photographers, who often refer to the 'speed' of their lenses, meaning that a high-speed lens can capture a lot of light and hence enable the production of sharp images of fast-moving objects, or produce acceptable images of static scenes at low light levels. The f-number is the ratio of the focal length of the optical system to the entrance pupil diameter; the lower the f-number the brighter the image. Both large eye size and low f-number are indeed found in the eyes of owls, suggesting that natural selection has operated to maximise sensitivity in the eyes of these birds.

Analysis of the optical system of Tawny Owl eyes shows that indeed the absolute size is disproportionately large compared to the weight of the bird, and the optical system has a relatively low f-number. Surprisingly, the overall size and focal lengths of the eyes of Tawny Owls and humans are very similar. However, they differ in the maximum brightness of the image which they produce. The f-numbers of human and owl eyes are 2.13 and 1.30 respectively. The result is that, when viewing the same scene, the image in an owl's eye is approximately 2.7 times brighter than in a human eye (it is the square of the f-numbers which must be compared).

That this difference is functionally significant comes from studies of the absolute visual sensitivity of Tawny Owls determined using behavioural training techniques (see Chapter 2). Compared to humans tested using the same apparatus, the absolute sensitivity of Tawny Owls is on average 2.2 higher than that of humans. This may not seem particularly impressive. However, compared to the strictly diurnal Rock Doves, owls are 100 times more sensitive. These findings indicate two things. First, that the difference between humans and owls with respect to absolute visual sensitivity is accounted for by the greater light-gathering capacity of the owls' eyes. That is, it is optics and not the retina that account for the gain in sensitivity. Secondly, because the f-number of dove eyes is very similar to that of human eyes, the doves' considerably lower sensitivity must be attributable to their retinas having lower sensitivity.

Visual sensitivity in context; in and out of the woods

The finding that natural selection has indeed maximised the visual sensitivity of owl eyes compared with other birds is not particularly surprising. What might be surprising is that fully dark-adapted owls are not significantly more sensitive that fully dark-adapted humans. This, however, raises questions about our own visual ecology rather than that of the owls. Clearly, owls are considerably more sensitive than diurnal birds.

While comparisons between owls and other species are interesting, perhaps the most important question is how the visual sensitivity of owls matches up to the light levels which occur naturally at night. Vision might be highly sensitive, but can it cope with all of the night-time conditions that an owl might face, especially beneath a woodland canopy?

Box 9.1 captures some of this light-level variability both inside a wood, beneath a tree canopy, and in open habitats. Comparing minimum light levels that occur inside and outside woods with the absolute maximum sensitivity of owl eyes shows that there will be occasions at night when light levels fall below the absolute visual threshold of an owl. This means that owls can be truly in the dark, unable to discern at close range even gross spatial details of objects around them. However, this failure to see anything will occur only beneath a woodland canopy. In open habitats, visual sensitivity is always sufficient to give owls (and humans, for that matter) some kind of vision at all naturally occurring night-time light levels. It is also important to note that under all conditions of natural illumination the sky will always be above the absolute threshold for vision. Even under conditions of starlight and thick cloud, owls can see branches, twigs, and leaves in silhouette against the sky.

Based on all of the above information, five general conclusions are warranted:

1. Absolute visual sensitivity in owls is close to the theoretical limit for a vertebrate eye – i.e. owl vision is just about as sensitive as it can be.

2. This sensitivity is not adequate to ensure that owls can gain visual information throughout the full range of light levels that occur naturally beneath a woodland canopy – i.e. owls will at times be in the dark, not able to detect visually any information about what is on the woodland floor below.

3. In open habitats vision of some sort is always possible, since light levels never fall below absolute threshold – i.e. outside woodlands some vision, even of the ground below, will always be possible.

4. The sky is always visible, which means that it can be a background against which objects can be seen in silhouette – i.e. by looking towards the sky there will always be some information available to guide orientation and flight, though this will often be restricted to gross detail.

5. The spatial resolution of owls at low light levels is superior to that of diurnal birds – i.e. despite all of the above four conclusions owls are better able to extract visual information at night than are diurnally active species that might live alongside them in the same habitat.

Overall these conclusions lead to an apparent paradox. An owl hunting outside a wood should always have sufficient light to see something. It is only beneath a tree canopy that gaining information through vision is sometimes impossible. Yet it is the most nocturnal owl species, those that complete all aspects of the life cycle between sunset and sunrise, that live beneath the woodland canopy. Owls that live outside woodlands in open habitats are the species more likely to be seen during the day or in twilight. Before it is possible to unravel this, it is worth considering briefly which other birds are also active at night and what they do at low light levels.

Nocturnally active birds

Strict nocturnality, as found in the woodland owls, is relatively rare among birds. However, many species show occasional nocturnal activity. Strict nocturnality is found only among non-passerines, most notably in some species of owls (Strigiformes), nightjars, Oilbirds, potoos and frogmouths (Caprimulgiformes), kiwi (Apterygiformes), and some species of parrots (Psittaciformes) (Figure 9.2).

Occasional nocturnal activity, in which some aspects of the annual activity cycle are completed after nightfall, is also primarily a feature of non-passerine species. There are, however, some notable examples of nocturnal activities among

Box 9.1 Natural light levels and vision: in and out of the woods

The light levels that animals can experience in natural environments are both highly variable and significantly different. Just ascribing light levels as those of 'daytime' or 'night-time' glosses over a wide range of light levels that vision has evolved to cope with. As shown in Box 2.4, natural illumination levels produced by the sun, moon, and stars vary over many orders of magnitude. Daylight illumination shows relatively little variation once the sun is above the horizon, although even these light levels change with season and latitude. However, the light levels of night-time can change significantly depending upon the presence, elevation, and phase of the moon.

These variations in light levels can amount to many million-fold and they make a significant difference to the amount of spatial detail that an animal can detect (see Box 2.3). However, there are additional factors which can alter the light levels experienced at any one place on successive days or experienced in

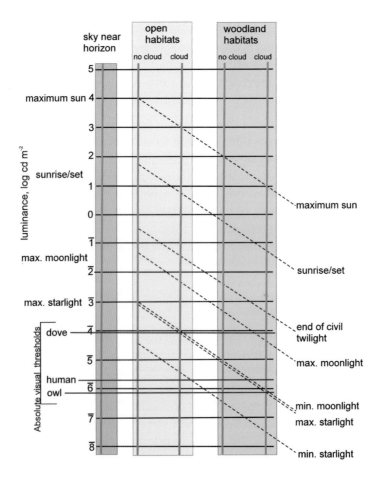

nearby places on the same day. These are the presence of cloud cover and the presence of a vegetation canopy. Cloud cover can reduce light levels by up to 10-fold, while just being under a tree canopy can reduce light levels 100-fold. The gloom of a forest compared with outside it is instantly experienced. These reductions of light levels due to cloud cover and vegetation canopies apply regardless of the ambient light source, whether it is sun, moon, or stars.

This figure captures some of the essential detail of how light levels change on the ground depending on the source of natural illumination (sun, moon, stars) and whether it is in the open or beneath a woodland canopy. The figure also shows the luminance of the sky near the horizon, as opposed to the ground upon which light is falling. This is important because it shows that even if it is too dark for an animal to see anything on the ground it is usually possible to see something in silhouette against the sky. It is always possible to see the sky. Even if there are thick clouds, the sky is always brighter than the ground, although mist and fog can obscure even this difference.

Along the vertical scale, the full ranges of luminance levels due to different natural light sources are shown: from maximum sun, through sunrise/sunset, maximum moonlight, to maximum starlight. The left-hand blue column shows the luminance of the sky without cloud. The yellow column shows the light levels experienced on the ground in open habitats, with and without thick cloud. The green column shows the same for beneath a woodland canopy, with and without cloud overhead.

The easiest way to read this is to follow the diagonal dotted lines that cross the *open habitats* and *woodland habitats* columns. For example, under maximum sun, light levels can vary by 3 log units (1000-fold) between being out in the open under clear skies and being in a woodland under cloud. Similar changes are seen across all sources of ambient light. The total light range falls by approximately 8 orders of magnitude (100 million-fold) from maximum sun and no cloud to minimum starlight. However, it is possible to experience a range of nearly 12 orders of magnitude (a million million-fold) between an open habitat in bright sunshine on a cloud-free day and beneath a woodland canopy under cloud cover and only starlight. Clearly, these are huge differences and alter very significantly what an animal (including ourselves) is able to see, especially the amount of spatial detail that can be detected.

The horizonal red lines give an indication of what these natural light levels might mean for three different species: Rock Doves, Tawny Owls, and humans. Each horizontal line indicates the minimum amount of light that each species can detect (the absolute visual threshold). Following the line for the dove from left to right, it can be seen that even if there is only starlight, the bird will be able to see the sky and can probably see something of the ground in open habitats. However, even cloud cover will decrease light levels below a dove's absolute threshold, and it will not be able to see anything. Entering a woodland means that a dove cannot see anything even under maximum moonlight if there is cloud cover.

Owls and humans, however, have significantly higher visual sensitivity than doves, and they should be able to always see the ground in open habitats, even under thick cloud cover, even if there is no moon. However, on entering a woodland, moonlight needs to be present in order to see something on the ground, and cloud cover results in light levels that are at the very edge of detectability. In other words, the sensitivity of owls and humans is always enough to see something except under conditions when the only source of illumination is starlight. The detail that can be detected close to absolute threshold will be very low, but something can always be seen.

passerines, although the behaviours involved are very restricted in scope. Among these is the night singing of some species. These include the well-known Common and Thrush Nightingales *Luscinia megarhynchos* and *L. luscinia* in Europe and central Asia, and Northern Mockingbirds *Mimus polyglottos* in North America and the Caribbean. In none of these species is singing restricted to night-time. Many species of passerines are known to undertake migratory flights at night, a phenomenon first confirmed through the use of the early radar systems which detected the movements of bird flocks at night at high altitudes. This behaviour has been confirmed many times by the use of sophisticated radar and radio-telemetry systems capable of detecting and following individual birds. The sensory problems faced by birds migrating at night are relatively simple, since flights take place at high altitudes well away from obstacles, with the birds departing on their flight around dusk and not usually landing until daylight hours. If birds do land at night they tend to stay put until light levels start to rise through the dawn.

FIGURE 9.2 Examples of strictly nocturnal birds. Top, examples of owls (Strigiformes) from the families Strigidae (owls) and Tytonidae (barn owls). The central block of four species are all from the Caprimuligiformes. Top left is a typical nightjar (Caprimulgidae), top right an Oilbird (Steatornithidae), below left a potoo (Nyctibiidae), and below right a frogmouth (Podargidae). Bottom row a kiwi (Apterygiformes) and a Kākāpō, a flightless parrot (Psittaciformes). (Photo of Western Barn Owl *Tyto alba* by Peter Trimming [CC BY 2.0]; Tawny Owl *Strix aluco* by the author; European Nightjar *Caprimulgus europaeus* by Dûrzan Cîrano [CC BY-SA 3.0]; Oilbird *Steatornis caripensis* by the author; Common Potoo *Nyctibius griseus* by Chiswick Chap [CC BY-SA 3.0]; Tawny Frogmouth *Podargus strigoides* by Benjamint444 [CC BY-SA 3.0]; Okarito Kiwi *Apteryx rowi* by Mark Anderson [CC BY-SA 4.0]; Kākāpō *Strigops habroptila* by Dianne Manson, Nationwideimages.co.nz.)

FIGURE 9.3 Occasionally nocturnal birds. These birds carry out most of the important aspects of their life cycle during daylight. However, all of these species may also be found foraging at night. They forage either in wetland habitats or in fresh and marine waters. Top row, Manx Shearwater *Puffinus puffinus* and Black Skimmer *Rynchops niger*. Middle row, Common Guillemot (Common Murre) *Uria aalge* and Northern Shoveler *Spatula clypeata*. Bottom row, Eurasian Oystercatcher *Haematopus ostralegus* and Double-crested Cormorant *Phalacrocorax auritus*. (Photo of Shearwater by Matt Witt [CC BY-SA 3.0], Skimmer by Gary Kramer USFWS [public domain], Guillemot by Boaworm [CC BY 3.0], Shoveler by Dick Daniels, carolinabirds.org [CC BY-SA 3.0], Oystercatcher by Andreas Trepte [CC BY-SA 2.5], Cormorant by Rodney Krey, USFWS [public domain].)

The occasional nocturnal activities of other bird species are usually associated with foraging (Figure 9.3). They are found in some species, or even just some populations, of the following taxa: shorebirds, skimmers, auks, and gulls (Charadriiformes), wildfowl (Anseriformes), cormorants (Suliformes), and penguins (Sphenisciformes), although many of these birds may commute to and within foraging areas at night and may also migrate at night. Many species of the procellariiform seabirds are active at night, primarily attending nest burrows under the cover of darkness. Some non-passerine species may give vocal displays at night, but not exclusively; most notable among these are the nocturnal vocalisations of crakes (Gruiformes) and cuckoos (Cuculiformes).

The key to understanding all of these nocturnal activities is to free ourselves from the idea that the birds must be behaving at night as we might expect them to do in daylight. It is clear that a bird which is active both during the day and at night does not follow the same patterns of behaviour round the clock, and nocturnal birds certainly do not exhibit the same behaviours as diurnal species. It is appropriate to think of these different behaviours, and the sensory information that guides them, as suites of adaptations which together provide 'solutions to nocturnality'.

The owls' solutions to nocturnality

Attractive as it may be to impute great wisdom to owls, this cannot, of course, be their solution. Even a wise bird needs information about its environment – and, as has been explained above, visual information is likely to be severely constrained in an owl's nocturnal environment. However, it does seem that knowledge of the environment plays a key role in nocturnality. Wisdom, or at least knowledge of a particular kind, can indeed be considered to have a role in explaining nocturnality in owls.

Hearing

The preceding sections of this chapter suggest an intriguing split between what might be possible guided by vision at night outside a wood, and what might be possible under a woodland canopy. Owl species which operate in open habitats may have sufficient visual information to guide their movements under practically all night-time conditions, especially as open habitats have very few small obstacles to avoid.

Short-eared Owls *Asio flammeus*, Barn Owls *Tyto alba*, and Snowy Owls *Bubo scandiacus* are examples of owls that live predominantly in open habitats. These open-habitat species specialise in taking a narrow range of small mammal prey from the ground. They may hunt for such prey during the day and in twilight, as well as at night. The principal problem that they face is not guiding their own mobility, but the detection of their prey, which is usually hidden beneath grass and leaf litter, sometimes under snow.

There is very good evidence that such prey detection is achieved primarily through the detection of sounds that are emitted as the prey disturb litter. The hearing of owls was described in Chapter 6, and its most notable feature is not its sensitivity but the high accuracy of sound localisation. It has been shown that sound cues are sufficient to allow various species of owls to capture moving prey guided solely by the sounds of leaf-litter rustles. Coupled with high visual sensitivity that provides gross spatial detail at low light levels, this use of auditory information must largely account for the foraging behaviour of owls that live in open habitats.

Fully nocturnal species, however, must do far more than just catch prey in darkness. They must complete all aspects of their life cycle at night, and furthermore they must be mobile in the structurally complex woodland habitats which they prefer to occupy.

Flight

Two adaptations of the flight of owls help to make such mobility possible. First, owls' wings are relatively large, resulting in high lift and slow flight speeds. These features can allow avoidance of obstacles detected at short range. Also, if the birds do collide with an obstacle the impact speed is low. This slow buoyant flight of owls thus plays a part in collision avoidance – or at least when impacts do occur, damage may not be serious. When owls pounce on their prey the feet are spread wide and the legs extended well forwards of the body, and they absorb the full force of any impact. In addition, the slow flight of owls is also very quiet. Not only is air displaced less vigorously by slow wing movements and low air speeds, owls also have particular adaptations to their feather structures (Figure 9.4). These reduce turbulence of the air flowing across the wing surface and result in near-silent flight, and this must function to reduce the probability that an owl will be detected as it flies towards its prey.

Hunting technique

Open-habitat owls, such as Barn Owls and Short-eared Owls, are able to hunt on the wing using slow flight and occasional hovering, effectively detecting their prey by sound from an 'aerial perch'. The woodland owls, on the other hand, must face many small rigid obstacles in their usual flight space, and it may be for this reason that they do not hunt on the wing. They use a perch-and-pounce technique and drop onto prey from fixed perches. This is the kind of task that owls were first shown to be capable of doing in total darkness. However, as discussed in Chapter 6, it is one thing to detect a sound, another to determine its direction, and yet another to determine its range.

Sound ranging can be performed in all birds and mammals only after extensive familiarisation with specific sounds and the ways that they are degraded under different environmental conditions. In Chapter 6 we saw that passerine birds may be particularly poor at both sound localisation and ranging. In owls, accurate localisation and ranging are possible, but only when the birds have had sufficient

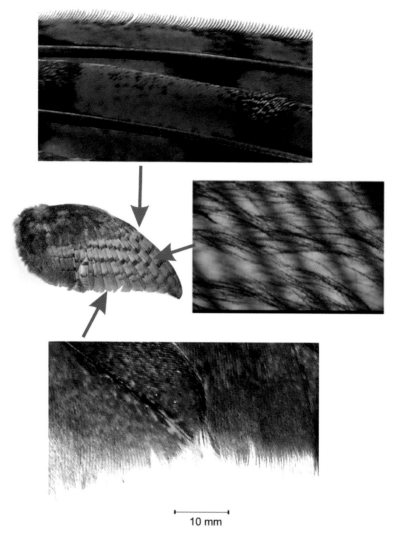

FIGURE 9.4 Features of owl wings and feathers that contribute to silent flight. The leading edges of primary feathers have a comb-like structure (top), the upper surface is covered in long hairs (middle), and the tips of the primary and secondary feathers have ragged soft ends (bottom). All of these features smooth the flow of air over the wing's upper surface and result in lower turbulence and less noise compared with the sharp edges and hard surfaces of the feathers of most birds. The overall shape of the wing gives a low aspect ratio (relatively deep compared with length) and the wing is also highly curved in cross-section. Combined, these features give an owl high lift at low airspeed, allowing it to fly at low speed without stalling. These features of the wing and its feathers produce the silent flight of owls. They serve not to disrupt an owl's own detection of sounds and also provide little warning of an owl's approach towards its prey. (Photos by Ennes Sarradj and Thomas Geyer.)

time to familiarise themselves with a hunting situation at low light levels. During this familiarisation period, which in captive birds may be a number of weeks, the birds are presumably learning both the positions of perches and the ranges of sounds heard from those perches. Thus, they become able to determine the range of sounds through repeated exposure. This need for familiarity with both sounds and perch positions introduces a particular behavioural adaptation which is a vital component to nocturnality in woodland owl species.

Knowledge and nocturnality

At the higher night-time light levels produced by moonlight, relatively fine details can be detected (Box 9.1). This could allow an owl in a woodland to become familiar with the finer spatial details of smaller branches, as well as with the structure of the larger trees that provide the overall framework of obstacles. This information, however, is quite specific to particular locations and will not generalise. Thus, the structure of a particular patch of woodland can become well known to an individual bird on the more brightly lit nights. This knowledge can be used at lower light levels when fine visual detail is no longer available but the larger structures, which define the framework for the location, can still be detected.

This introduces the idea that knowledge of various kinds plays a clear role in the nocturnal behaviour of owls. Not just knowledge of sound degradation with distance allowing accurate ranging, but also knowledge of the spatial structure of a particular location. This knowledge is acquired when light levels are high and applied when light levels are low. This means that rather minimal or partial spatial information can be used to guide mobility at the lowest light levels.

There is good evidence to support this idea. It comes from the high degree of territoriality exhibited by woodland owls, something which is not found in owls of open habitats. Tawny Owls are, in fact, among the most sedentary of all bird species studied to date. Basically, Tawny Owls do not go anywhere. Once settled, all of their lives are lived out in relatively small patches of woodland habitat which may be no more than 5 hectares. Possession of a territory is crucial for annual survival. Tawny Owls have an average life span of 4+ years, and many live for more than 20 years.

Once a young bird has established a territory in the first few months of its life, annual survival can be as high as 85%, but failure to establish a territory results in an equally high probability of death. Both sexes defend their own individual territories, but they overlap to some extent in breeding pairs. Interestingly, territory boundaries out-survive their owners, suggesting that occupancy on either side of the boundary can change due to the death of a holder but an adjacent holder does not expand into the vacant territory. The reason for this is that a neighbour cannot readily move in and exploit the resources of an adjacent vacant patch because it lacks sufficient knowledge of the structure of the patch. Thus, a new bird may take over a vacant territory and learn its existing boundaries by being driven away by adjacent territory owners, who stick rigidly within the patches that they know.

Such a high degree of territoriality is not a feature of owl species which live in open habitats. These birds may hold a territory for an extended period, but it is not so essential for survival as it is for the woodland owls. Open-habitat owls will defend the resources of the territory during breeding, but outside of the breeding period they may wander, and some species, such as Short-eared Owls and Barn Owls, periodically range widely, even crossing stretches of sea, for example between continental Europe and the British Isles. These birds do not have regular migratory patterns but seem to undertake movements away from breeding areas primarily in response to shortages of their preferred prey types, which are small mammals that are prone to population cycles.

Woodland owls do not follow such patterns, and they are far less specialised in their diet. They take a wide range of prey, including earthworms and beetles that would seem of suboptimal size compared with the small mammals which typically make up the bulk of their diets. This dietary breadth results from the fact that woodland owls simply cannot move away from their familiar patch in pursuit of more optimal prey. In effect woodland owls have to sit tight and broaden their diet rather than move to seek out more optimal size prey in novel locations. The open-habitat owls do not experience this constraint, since their prey capture and mobility does not depend upon such detailed knowledge of a specific location.

Owls in summary

The nocturnality of owl species is based on two different strategies.

First, there are the owls of open habitats. These rarely, perhaps never, experience the lowest range of naturally occurring light levels when they are abroad. There is always sufficient light to give them the benefit of visual guidance, albeit with relatively low spatial resolution. Their preferred habitats are spatially uncluttered, and large obstacles can be detected from distances that are sufficient to avoid collisions when flying.

Second, there are owls which live under a woodland canopy. They face a more exacting task in that light levels on the forest floor may frequently fall below their threshold for vision. Under many circumstances fine spatial details cannot be detected even at close range. The solution for these birds is to build knowledge of the spatial structure of a particular location, which becomes the place in which they live the whole of their lives. This knowledge, their 'wisdom', allows resident birds to interpret the limited spatial information that is available to them under the lowest light levels.

In both open habitats and in woodlands, it is clear that vision alone is not sufficient for the detection and capture of prey. There is no super-sensory, gee-whiz solution to owl nocturnality. Living nocturnal lives in both open and woodland habitats is possible because of a combination of sensory information and behavioural adaptations. The movements of potential prey through vegetation produce sounds which owls are able to locate accurately, in both direction and range. Vision provides spatial information, but only about relatively gross detail, and sometimes light levels are below threshold and so no spatial information can be available. Thus, in both open-habitat and woodland owls, their senses of vision and hearing work in tandem, complementing each other and providing sufficient information to allow prey capture and mobility at night-time light levels. To exploit the resources of a woodland habitat at night requires the additional behavioural adaptation of a highly sedentary life. This facilitates the accumulation of knowledge that allows minimal spatial cues to be successfully interpreted. In truth, 'the wise old owl' probably has little option other than to 'sit in an oak'.

The Oilbird's solution to nocturnality

While owls are widely regarded as the archetypical nocturnal bird, they are not the most nocturnal. That accolade belongs to Oilbirds (Figure 9.5). The majority of Oilbirds probably never experience daylight in their entire lives. Oilbirds were first described scientifically in 1817 by the great German explorer-scientist Alexander von Humboldt. His descriptions of their biology, and the specimens that he collected, baffled taxonomists, so much so that Oilbirds were first placed in an avian order of their own. An order with just a single species suggested that these birds were so unique that they had no close living relatives. Today, Oilbirds are placed within the Caprimulgiformes (nightjars and their allies) but in a family of their own, Steatornithidae.

Von Humboldt came across Oilbirds in 1800, towards the end of his first expedition through the tropical rainforest of South America. He found them near the mission station of Caripe, now a small town in Venezuela. The mission station's name became part of the scientific name of these birds, *Steatornis caripensis*, which translates as 'the fat bird of Caripe'. The name is derived from the nestlings, which lay down a lot of fat and were traditionally harvested and rendered to provide oil

FIGURE 9.5 Oilbirds *Steatornis caripensis*. The close-up photograph of the head shows two of the key sensory systems of these birds: the prominent and long rictal bristles around the mouth, and the large eyes whose sensitivity is thought to be close to the limit for a camera eye in a terrestrial vertebrate. Two Oilbirds captured by flash photography sit on a ledge deep in the interior of a cave where no light penetrates. (Photo of two Oilbirds by The Lilac Breasted Roller [CC BY-2.0].)

for cooking and light. Caripe has been renamed Caripe del Guácharo. *Guácharo*, the Spanish name for the species, is an onomatopoeic rendition of their call. The renaming of the town reflects the minor tourist industry that has grown up around these birds, and visitors to the town today are greeted on main street with a 5-metre-wingspan model of an Oilbird. Near the town is the Cueva del Guácharo National Park, which is devoted to the conservation of these birds.

Oilbirds roost and breed communally in caves at depths where no light penetrates, and they emerge from their caves at dusk and return before dawn. Although some birds may roost in trees during the day, the majority of birds never see daylight. At night they forage in the rainforest for fruit, which makes them unique among the Caprimulgiformes. Clearly these birds face some interesting sensory challenges, for not only do they have to orient themselves and control flight in the total darkness of cave interiors, but they have to guide their foraging for fruits at night.

Vision in Oilbirds

Oilbirds have eyes that are relatively large with respect to the bird's body mass, but they are little more than half the length of the eyes of Tawny Owls (16 mm, compared with the owl's 28 mm). As in owls, the eyes appear to bulge from the skull, but they are in fact positioned within orbits. There is good evidence that these eyes are among the most sensitive in terrestrial vertebrates and are very close to the theoretical limits of visual sensitivity. This is achieved by the eye's optics having a very low f-number combined with a retina that is dominated by rod photoreceptor cells. The f-number of Oilbird eyes is 1.07, resulting in Oilbirds having an image brightness 1.5 times higher than in Tawny Owls and nearly 4 times brighter than in human eyes.

Analysis of this bright retinal image is carried out by a retina dominated by rod photoreceptors, and there is evidence that in some parts of the retina the rods are arranged in multiple layers. This is a mechanism that enhances sensitivity by increasing the probability that photons in the retinal image will be intercepted. Efficient as this mechanism is for enhancing sensitivity, it is, however, likely to decrease spatial resolution. Tiered retinas of this kind are found in some deep-sea fish which live permanently at extremely low light levels, locations where sunlight never penetrates, and the only light comes from the photophores of animals. The low f-number of Oilbird eyes, combined with a highly sensitive retina, means that their eyes are pushing at the theoretical limits of visual sensitivity for vertebrate eyes, at least for terrestrially active species.

From the perspective of vision, Oilbirds appear well adapted to the demands of life at low light levels. However, as in owls, vision cannot be the sole answer to their mobility or their foraging at night. Deep inside caves where Oilbirds nest and roost, no photons of light penetrate, there is simply no light to be seen. There can be no residual vision providing even gross spatial cues.

Hearing in Oilbirds

The hearing of Oilbirds was touched on in Chapter 6. In the 'Active sonar: echolocation' section of that chapter we saw how hearing in this species has evolved to enable spatial information to be detected using echolocation (active sonar). There is no evidence that Oilbird hearing differs from hearing in other birds in either sensitivity or frequency range. However, the birds are capable of emitting pulses of sounds, clicks, whose echo is detected and analysed to provide information about the environment. The spatial resolution of this system is poor, enabling the detection only of large obstacles at close range. However, it must provide sufficient resolution to enable birds to locate cave walls, other birds, and possibly nest and roosting ledges, within the caves' totally dark interiors (Figure 9.6).

As in owls, the wings of Oilbirds are large, resulting in a low wing loading and therefore slow flight; Oilbirds are capable of momentary hovering in the still air of a cave. These features must reduce the probability of damage should collisions occur. Thus, as an Oilbird enters deep into its cave spatial guidance must switch from visual to auditory-based spatial information. Outside the cave the birds do not echolocate; they fly in silence.

FIGURE 9.6 Active sonar in Oilbirds. An Oilbird is depicted hovering before a flat disc suspended in its flight path within the lightless interior of a cave. The bird emits high-intensity pulses of sound (red) of relatively low frequency. The pulses of sound are reflected back to the bird from the disc (blue), and these reflected sounds allow the Oilbird to detect the obstacle. However, only relatively large objects can be detected, and only at close range. The bird's large wings provide a low wing loading which allows it to fly slowly and momentarily hover when detecting obstacles or the walls of the cave.

When the birds leave the cave at dusk, visual guidance takes over. The actual task may not be particularly taxing and is akin to the challenges faced by open-habitat owls, for although these are birds of tropical rainforest habitats they fly above the canopy. In effect they fly in an open habitat devoid of fine spatial obstacles. As in the owls of open habitats, light levels encountered above the tree canopy will always exceed visual threshold. This will allow visual orientation to large obstacles on the darkest of nights and probably also the detection of relatively fine spatial detail under moonlit conditions.

Olfaction and tactile cues in Oilbirds

To guide their foraging as they fly over the canopy, Oilbirds switch to a third telereceptive sense, olfaction. Although there is no detailed experimental work on olfaction in Oilbirds, good observational evidence indicates that they are using olfaction to at least detect locations where ripe fruits are concentrated. This is akin to the use of olfaction by seabirds to find foraging locations in the ocean (Chapter 6).

Oilbirds may be envisaged as flying over the canopy searching for odour plumes from trees bearing ripe fruits. Once detected, this would allow the birds to enter the upper layers of the canopy and perhaps seek out individual fruits using not only olfactory and probably visual information, but also tactile cues.

Like other nightjar species, Oilbirds have prominent rictal bristles around the mouth area (Figure 9.5). It seems likely that these have a tactile function and could guide Oilbirds' mouths towards individual items. Such tactile information would also facilitate close contact between individuals when on roosting and nest ledges in the complete darkness of caves. Information from taste is also likely to come into play for the identification of edible fruits.

Oilbirds in summary

Oilbirds may be an extraordinary species, but they do point up some general lessons. They show what can be achieved using both sensory information and behavioural adaptations when light levels are low, or even absent. Natural selection has driven high sensitivity, but this will be at the cost of low resolution. It seems clear that Oilbirds must rely on gross spatial information to guide their behaviour at all times. Fine-grained spatial information is never available to them.

Although both vision and echolocation play key roles in providing spatial information to guide their behaviour, these two senses do not work together. Echolocation takes over when vision is no longer viable in the cave, and vision is the default sense when there is natural light available outside the cave. Effective as these two information sources are, a third telereceptive sense, olfaction, is employed in the guidance of foraging. The intimate senses, touch and taste, provide key information for the selection and acquisition of food items.

Oilbirds are certainly extraordinary. They challenge commonly held ideas of what birds do and how they do it, and bring a different perspective to the idea of a 'bird's-eye view'. However, alongside the owls they provide a particularly clear illustration of how it is not always necessary for flying birds to have available to them highly detailed spatial information of the world about them. Both owls and Oilbirds show us that whether detailed spatial information is necessary depends on specific environmental conditions and specific behaviours. Oilbirds also show that the senses of vision, hearing, and olfaction can provide spatial information at different scales and can complement each other in providing information to guide key tasks.

The kiwi's solution to nocturnality

As with owls and Oilbirds, the sensory ecology of kiwi provides a fascinating narrative. Five species of kiwi are currently recognised, all endemic to New Zealand. They have become symbols of both the nation and its unique avifauna which evolved in the absence of terrestrial mammals over a period of 80 million years. The name kiwi is from the Māori language and is onomatopoeic, based on one of the kiwi's common calls. As a Māori word it does not take an s to indicate plural. Whether reference is to one or many individuals, or to one or many species, must be understood from the context, all are kiwi.

Kiwi are both nocturnal and flightless. Therefore, some of the challenges faced by owls and Oilbirds are ruled out. However, feeding exclusively on the forest floor presents other challenges because their prey is always hidden from sight, and this has led their foraging to become dependent upon information about objects that are very close. The major telereceptive senses have declined in importance in favour of the intimate senses.

On first consideration the senses present a paradox. Asked to guess which senses might be important to a nocturnal flightless species, it would be reasonable to suggest that vision should be important. After all, as explained in Chapter 3, to achieve high sensitivity eyes of absolutely large dimensions and with a low f-number are necessary. Being flightless, kiwi would not seem to face the kinds of low weight requirements faced by flying birds. Eyes could be huge, capable of catching a high number of photons from the light that penetrates to the forest floor. If size and weight are not restrictions on eye design, why not gain high sensitivity?

Contrary to this expectation kiwi have small eyes (Figure 9.7). Their axial length is only 7 mm, smaller than those of Common Starlings. In proportion to body mass, kiwi eyes are tiny. Furthermore, flightlessness in other bird species has tended to favour the evolution of large eyes. Among all birds, the eyes of the flightless ostriches, rheas, cassowaries and Emus, and penguins are among the largest, suggesting that flightlessness removed an important constraint upon eye size in these birds. So what has happened in the evolution of kiwi to have led to such small eyes, when large size is readily predicted?

FIGURE 9.7 Kiwi. The close-up photograph of the head shows that the eyes are small, especially when compared with those of an Oilbird (Figure 9.5). However, like Oilbirds, kiwi also have long and prominent rictal bristles around the mouth. The photograph of a kiwi foraging on a forest floor captured by flash photography also emphasises the small size of the eyes but also shows the bird placing its bill close to the forest floor and probably using the sense of smell to locate items, with nostril openings just behind the bill tip (see also Figure 7.4). (Close-up of a North Island Brown Kiwi *Apteryx mantelli* by the author, foraging Southern Brown Kiwi *A. australis* by Glen Fergus [CC BY-SA 2.5].)

The initial conclusion must be that kiwi cannot be relying upon vision as the primary source of information to guide their nocturnal behaviour. But this raises the question of why give up on visual information when other nocturnal birds have eyes that have evolved to provide high sensitivity?

Olfaction, touch, and hearing

The use of olfactory information in the foraging behaviour of kiwi was first demonstrated experimentally more than half a century ago by Bernice Wenzel. This was backed up by work on the relative sizes of olfactory bulbs in birds which showed that in kiwi a large proportion of the brain is devoted to the analysis of olfactory information. As described in Chapter 6, kiwi are unique among birds in having their nostrils at their bill tips. This arrangement can clearly function as a means for olfactory sampling at a relatively fine spatial scale, especially of a surface on which the bill is placed, or just below the surface when the bill is probed into it.

The behaviour of foraging kiwi has been described by Hugh Robertson as 'walking slowly along tapping the ground and when prey is detected they probe their bill into the leaf litter or a rotten log; occasionally plunge their bill deep into the ground'. It seems likely that the 'tapping' behaviour could be involved in the collection of olfactory, as well as tactile, information.

Kiwi have a well-developed bill-tip organ (Chapter 7), and experimental work has shown that detailed tactile information is gained using this organ. The bill-tip organ may also be capable of detecting items buried in wet substrates some distance from the tip, using 'remote touch'. Furthermore, around the base of the bill there is an array of long rictal bristles, more extensive than those found in Oilbirds, which probably function as remote tactile receptors.

Kiwi are highly vocal, and like many birds use auditory signals to control social interactions between pairs and to make declarations in territorial disputes. There is evidence of some specialisation in the anatomy of the auditory system of the brain, although to date there is no experimental evidence on auditory abilities. The ability of kiwi to determine the direction and range of sounds is unknown. Kiwi lack elaborate outer ear structures of the kind found in owls. However, their skulls are rather broad and solid compared with passerine species so their ability to locate sounds may be more acute than passerines, since inter-ear differences in sound intensity and the time of arrival of sounds may be greater and provide better directional cues (Chapter 6).

The sensory world of kiwi

The sensory world of kiwi is very different to the worlds of other nocturnal bird species. Pedestrian locomotion across a dimly lit forest floor, often using regular tracks, seems to pose simpler challenges than those faced by owls and Oilbirds in their night-time flights. The kiwi diet is mainly composed of relatively immobile invertebrates, whose detection and identification can clearly be achieved by a combination of tactile and olfactory cues. Furthermore, the prey-detecting mechanisms are incorporated into the very tool, the bill, which is used to explore and manipulate the environment. This is a neat combination that provides a self-guided tool with the sensory detectors right at the very point at which it operates. There is no need to estimate direction and distance to a food item, since when it is discovered it is right at the bill tip. The business of prey capture is more or less the same as prey detection.

But what about general mobility within the forest floor? Surely vision ought to be able to help with general orientation and guidance while foraging or inter-acting with another kiwi? The visual abilities of kiwi have not been investigated systematically, but while their small eyes should provide some spatial information, the resolution is likely to be much lower than in owls, for example.

As in the woodland owls, a cognitive component, knowledge of specific locations, may also be vital to explain the nocturnal mobility of kiwi. This can be achieved through these birds' long-term residence in the same patch of habitat, which is defended against neighbours. This territorial system may parallel that of Tawny Owls in that once established in a territory the birds never leave that area for the rest of their lives. It can be surmised that through this long-term residence detailed knowledge of a habitat's structure can be gained even from the minimal spatial cues that are available.

Like owls, kiwi are long-lived birds; in fact they probably live longer than owls on average. Southern Brown Kiwi *Apteryx australis* have been recorded as living up to 35 years in captivity and at least 20 years in the wild. Furthermore, it has been shown that their populations are extremely subdivided with many so-called 'cryptic species', suggesting that kiwi are highly sedentary with very low dispersal from natal areas. In other words, kiwi stay put and live long, and this may be the key behavioural component to their nocturnal existence, just as it plays a key role in the lives of woodland owls.

Are kiwi giving up on vision?

Although it is possible to make this sensory ecology narrative about the nocturnal behaviour of kiwi, an intriguing question remains. Why is vision of such little importance to kiwi? Owl-sized eyes could have evolved in kiwi and provided resolution that would have made moving around the forest less challenging. Are the small eyes of kiwi species today a legacy from the time when kiwi were diurnally active and their ancestors arrived in New Zealand? Did the eyes simply stay small as the birds became more nocturnal and started relying on non-visual information?

It has been argued that this is not the case. The reduced reliance upon visual information is more likely to be a derived characteristic in kiwi, meaning that the early ancestors of kiwi, which were diurnal and capable of flight, lost their reliance on vision as they became nocturnal and flightless. This downgrading of vision in kiwi is probably an example of adaptive *regressive evolution*. At some point in the evolution of kiwi, natural selection favoured forgoing visual information in favour of other sensory information. In other flightless birds, Emus and cassowaries, which share a common ancestor with kiwi, vision has remained important, but in kiwi it has been downgraded.

The ecological circumstances favouring this downgrading of vision in kiwi can only be speculated on. Regressive evolution of visual systems has been described in both vertebrate and invertebrate animals, but these examples have involved a complete loss of vision following colonisation of subterranean habitats devoid of all light. In kiwi, complete regression of the eye has not occurred.

Probably the key factor that allowed the regression of vision has been the absence of mammalian predators throughout the whole period of kiwi evolution in New Zealand. Avian predators (Haast's Eagle, Eyles's Harrier) were present during this time, and certainly Haast's Eagles were large enough to take birds the size of modern kiwi. It could well be that it was predator pressure from these birds' daytime hunting that set in train the evolution of nocturnality in kiwi.

Not only are the eyes of kiwi small, their visual fields are also very small. The field of a single eye is the narrowest so far described in any bird, and the total visual field and the binocular field are the smallest of all birds determined to date. There is a very narrow area of binocular overlap and a very wide blind area behind the head (Figure 9.8) which leaves them unable to detect predators visually from a large segment of space around them. Thus, it seems that an absence of nocturnal predators is likely to have been a key factor in driving the regression of kiwi vision.

But why give up on vision? Surely it would always be useful to see something? This, however, is an argument from a human perspective, the perspective of a visually guided animal for whom vision is an important factor in daily survival. There is a cost to vision, a metabolic cost. As pointed out in Chapter 2, eyes are expensive organs to run and vision takes up a lot of processing power and hence

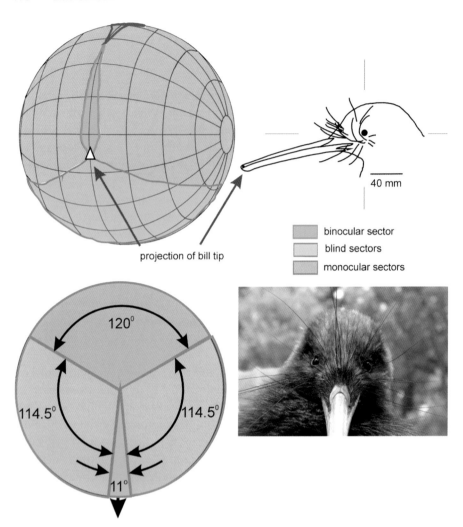

projection of bill tip

40 mm

- binocular sector
- blind sectors
- monocular sectors

120°

114.5°   114.5°

11°

FIGURE 9.8 The visual fields of kiwi. These visual fields are quite unlike those described in any other birds species (see various examples in Chapter 5). The most interesting features are the very small size of the region of binocular overlap, depicted in green, the small size of the field of each eye, and the extensive blind region behind the head (depicted in blue). The binocular field is both narrow and of short vertical extent. The lower diagram shows a horizontal section through the field in the plane from which the bird is viewed in the photograph. It shows that the width of the field of a single eye is 125°, much narrower than found in all other bird species investigated to date. It is also notable that a Kiwi cannot see its own bill tip; this is quite unlike the situation in birds which use vision to guide the movement of their bill towards a target. All of these features strongly suggest that, unlike in other nocturnal birds, vision is of very little importance to kiwi, and that their vision has undergone regressive evolution.

energy in the brain. It seems likely that the regressive evolution of kiwi vision is the result of the trade-off between the requirement for an eye large enough to gain useful spatial information at low light levels, and the metabolic costs of maintaining such large eyes (and the costs of extracting and processing information from them). In kiwi the absence of mammalian predators may have tipped this balance so that the costs of vision against the information that other sensory systems could provide led to these birds becoming primarily guided by non-visual cues.

This analysis can account well for the different kinds of information that kiwi employ in a complementary manner to control their nocturnal behaviour. It also provides another good example of how the use of different sensory information is tuned to the sensory challenges of particular environments, and to the conduct of particular tasks.

Lessons from owls, Oilbirds, and kiwi

These narratives on owls, Oilbirds, and kiwi show that a fully nocturnal life depends on a suite of information from different senses, plus some specific behavioural adaptations. Behaviour and different sensory information complement each other in different and complex ways in each of the examples. There is no single or optimal way of being nocturnal; it depends on the diet, habitat, and evolutionary history of the species. So there are no simple answers, but there are general principles at work. One key finding is that many species must be completing their nocturnal behaviours using only partial information, partial at least when compared with what we might expect birds to have available to them during the light.

Chapter 10

Other birds of the night: parrots to passerines

There are other bird species which are totally nocturnal, but unfortunately their sensory capacities and behavioural adaptations have not been subject to detailed investigations. In the case of some species, even their general natural history is only poorly understood. For such species, the construction of explanatory narratives of the kind presented in the previous chapter is not possible. Each of these species would make very rewarding subjects of study using a sensory ecology approach. In addition to these other truly nocturnal bird species, there are many which can be active at night for the conduct of specific behaviours, often at particular times in their annual life cycles. A brief survey of some other nocturnal species and of the occasional nocturnal behaviours of otherwise diurnal species follows.

Kākāpō

There are two species of nocturnal parrots: Kākāpō *Strigops habroptila* and Night Parrots *Pezoporus occidentalis*. Both are extremely rare with populations of probably less than 200. Night Parrots are found in a distinctive type of grassland, *Spinifex*, in dry areas of Australia. They were once thought to be extinct but were reported to have been rediscovered in 2003, and after extensive searches a live specimen was caught in April 2015; others have been tracked down since. Their ecology, behaviour, and anatomy are hardly known and so unfortunately it would be inadvisable to conjecture about their sensory ecology. Kākāpō, on the other hand, are better known. They have featured in many wildlife films, in which they are usually presented as a conservation success story, a species brought back from the brink. Their future is still uncertain, and they remain Critically Endangered, but because of this their breeding biology has been intensively studied – and this has thrown some light on their sensory ecology.

Kākāpō, like kiwi, rely on resources found in predator-free native forest habitats in New Zealand (Figure 10.1). They are flightless and wholly nocturnal. Unlike kiwi, they live on an exclusively vegetarian diet. Depending on the season they forage for different fruits, berries, nuts, seeds, green shoots, leaf buds, roots, rhizomes,

FIGURE 10.1 A Kākāpō foraging on the floor of a native forest habitat on Maud Island, one of the small number of predator-free protected islands used in this species' population recovery programme. The eyes are smaller than in owls and Oilbirds but larger than those of kiwi. Rictal bristles around the mouth are prominent but not as well developed as in Oilbirds and kiwi (compare with Figures 9.5 and 9.7). (Photo by New Zealand Department of Conservation [CC BY 2.0].)

tubers, bark, stems, and fungi. They even eat fern fronds and club mosses. All of these food sources have generally low nutritive value, and large quantities have to be eaten.

Kākāpō are described as 'living life in the slow lane' in that they do everything more slowly than most birds. They live 58 years on average and up to almost 90 years has been recorded, and reproduction is intermittent, occurring only in years when certain foods are in abundance.

Through the lens of sensory ecology, the life of a Kākāpō appears not to be as challenging as those of other nocturnal bird species. Like the kiwi they evolved in an environment free of natural mammalian predators, so vigilance for predatory detection is unlikely to have driven the evolution of their senses. Like the woodland owls and kiwi, they live under a forest canopy where light levels can be very low at night. In common with those species, Kākāpō are also highly sedentary, living long and well-regulated lives within their territory, so much so that they make defined pathways. This clearly suggests that Kākāpō parallel the kiwi and the woodland owls in having the opportunity to gain detailed knowledge of particular locations, and that this knowledge can be built up through very long-term residence and regular movement patterns under the full range of naturally occurring night-time light levels. Thus there may well be an important cognitive component to their nocturnality, and because these birds move slowly there is little danger from collisions with unpredicted obstacles.

The eyes of Kākāpō are relatively large in terms of axial length and corneal diameter. They are, however, well within the range of other parrots and scale with head size and body mass. This suggests that their eye size is not driven by optical adaptations that provide increased sensitivity. Nothing is known about their retinas except that photoreceptors seem similar to those of other parrots. There is some evidence that Kākāpō have a decreased reliance upon vision in that the proportion of the brain involved in analysis of visual information is reduced compared with diurnally active parrot species. This suggests that spatial resolution may be lower than in diurnal parrots of similar eye size and body mass, but there is no evidence that the Kākāpō's vision has undergone regressive evolution, as found in kiwi.

Kākāpō clearly use auditory signals to regulate their social behaviour, especially during breeding – but of course hearing will not play a role in their foraging for plant matter. Like other parrots it can be assumed that Kākāpō have a bill-tip organ, but this has never been investigated. Such a clustering of tactile receptors would function in the seizure and manipulation of objects regardless of light level.

Olfaction may be an important source of information for Kākāpō. Their brain has a relatively large olfactory bulb, and they also produce a distinctively sweet-smelling feather odour which could be used as a sociochemical signal (Chapter 6). There is also evidence that Kākāpō use olfactory information when foraging. This comes from studies with a tame bird trained to go to a food hopper. It was found that the bird could find food when it was hidden from sight. This evidence is not robust, but further work on olfactory signals could prove very interesting and might have a role in their conservation management.

The sensory ecology of nocturnality in Kākāpō thus seems to throw up both parallels and differences with the sensory ecology of the other obligate nocturnal birds. Perhaps surprisingly, Kākāpō do not seem to show evidence of particular sensory specialisations of the kind found in kiwi, owls, and Oilbirds. However, in Kākāpō and these other species, nocturnality depends upon a combination of sensory information and specific behaviours. Perhaps the behavioural key for Kākāpō is their slow pace of life, territoriality, and longevity: these allow the birds to gain knowledge of their specific environment at close proximity. Kākāpō may explore their environment using low-resolution information from vision and tactile information from their bill tips, possibly combined with specific information from olfaction.

Nightjars and their allies

The order Caprimulgiformes contains about 120 species which are active at night and/or twilight. With the exception of Oilbirds, their senses have not been investigated in any great detail so understanding of their sensory ecology is not on a very firm footing. However, it is possible to piece together some general ideas about the information that they have available to guide their nocturnal behaviours. There are three major families in the Caprimulgiformes, and they present a similar division as is found between the woodland and open-habitat owls; this division may be key to understanding their sensory ecology. Nightjars (Caprimulgidae) are birds of open habitats and exploiters of insects taken from the open airspace. Frogmouths (Podargidae) and potoos (Nyctibiidae) are birds which forage primarily beneath woodland canopies for insects.

Nightjars

Nightjars forage at night, mainly in open habitats or above a vegetation canopy. They take larger flying insects, especially the moths and beetles which start to fly as the evening progresses, but smaller insects are also common in their diets. The foraging techniques and the specialised anatomy of the mouth have been the subject of a number of investigations, and these give a valuable clue to the importance of sensory information for these birds' foraging.

There is good evidence that nightjars capture prey by 'open-billed trawling' through dense clouds of insects, but they may also pursue larger individual prey items. Two specialised mouth structures, and a possible touch-sensitive region inside the mouth, suggest that nightjars may be exceptionally well adapted for the capture of insects by trawling, but when trawling they may be flying almost blind.

Rictal bristles grow from around the margin of the upper bill (Figure 10.2). As discussed above, rictal bristles probably have an important role in guiding the nocturnal behaviour of Oilbirds and kiwi. In nightjars the bristles are denser and probably help to direct insects towards the open mouth, thus enlarging the area that is trawled beyond the margins of the mouth. The bristles probably also provide tactile information that could indicate when insects have been intercepted.

FIGURE 10.2 European Nightjar *Caprimulgus europaeus*. As in other species in this family (Caprimulgidae), the small bill that is seen when the mouth is closed gives way to a very wide gape as the bill opens. This is achieved by special joints in the mandible (red arrow) that spread the bill wide open as it is lowered. A fully opened gape is approximately circular. The upper bill is surrounded by rictal bristles which serve to increase the cross-section of the area swept by the bird as it trawls airspace for insects, and which may also indicate when an insect is intercepted. The roof of the mouth is highly vascularised and may contain touch receptors that detect impact with insects as the bird trawls almost blindly. (Photos by Dûrzan Cîrano [CC BY-SA 3.0]; Raymond Galea, Birding in Malta.)

The open gape of nightjars is surprisingly large, much larger than might be supposed from what is seen of the bill when the mouth is closed. This large gape is achieved because of an arrangement of joints, flexible bones, and musculature involving the articulation of the mandible. This unique anatomy spreads the sides of the mandible wide apart and results in a nearly circular cross-section to the open mouth. However, with the gape fully open, both the maxilla and the rictal bristles may obstruct the eyes, stopping the bird from seeing clearly forwards.

The roof of the mouth (palate) of nightjars has been reported to contain a unique sensory structure directly linked to open-billed trawling. In the majority of bird species, the palate is lined by a horny sheath which is relatively tough and apparently insensitive to touch. In nightjars the horny sheath is absent and in its place is a highly vascularised membrane, which is bright red in colour because of the high number of blood vessels close to its surface (Figure 10.2). This was first described by Graham Cowles of London's Natural History Museum over 50 years ago. He conjectured that this membrane 'would be very sensitive and easily stimulated by an insect striking the surface, enabling the bird to react quickly to the prey while in flight'. To really act as a sensory structure, we would predict that it is rich in touch-sensitive receptors, not just blood vessels. There are no reports of this, but no one has ever looked. If found, this would be a fascinating specialisation of touch sensitivity inside the mouth, somewhat akin to the bill-tip organs in ducks and shorebirds which allow the detection of prey by tactile cues alone (Chapter 7).

It seems, therefore, that nightjars have two, possibly three, anatomical features which could act to aid the capture of insects without the prey even being seen. The picture emerges of nightjars trawling open airspaces, their large mouths wide open, with a capture area enlarged by rictal bristles, possibly acting as a touch-sensitive net. Another touch-sensitive mechanism inside the mouth indicates that insects have struck the palate, and the mouth can then be snapped shut and the items swallowed.

Little is known about the visual capabilities of any species of nightjar, but what evidence there is suggests that vision is unlikely to provide sufficient spatial resolution to allow it to take a role in foraging for small insects. First, measurements of visual fields in Pauraques show that they have a small degree of binocular overlap. Secondly, the eyes of some nightjars, including Pauraques and Common Poorwills, have been shown to contain a tapetum (Figure 10.3).

Nightjars are the only birds with eyes that contain a tapetum similar to those found in some mammals and fish. In nightjars the tapetum (or tapetum lucidum) is a structure made up of very small reflective spheres of lipid behind the photoreceptor layer in the retina. It functions to increase the chance that light will be intercepted by the photoreceptors by reflecting light back out of the eye, so if a photon is not caught initially there is a second chance to capture it. It is found in mammals such as cats and dogs, though not in primates, and not in most birds. The tapetum produces a diffuse reflection that gives rise to eye-shine when the eyes are caught in a bright light at night.

FIGURE 10.3 A Common Poorwill *Phalaenoptilus nuttallii* (top) and a Pauraque *Nyctidromus albicollis* (bottom): two species of nightjar which have been shown to have tapeta in their retinas. The photo of the Poorwill clearly shows eye-shine. This is light from the camera flash that is reflected back out of the eye from the tapetum that lies behind the retina. Note that this is not the same as the red eye seen in flash photographs of human eyes, which is produced by light reflected from the front surface of the retina, which is coloured by the blood vessels lying on the surface of the retina. Birds do not have blood vessels across the surface of their retina, so the red colour comes from the tapetum. (Photo of Poorwill by Ian Routley, Pauraque by Karin Schneeberger [CC BY-SA 3.0].)

This results in an increase in the absolute sensitivity of an eye, but it almost certainly reduces spatial resolution. This may not be disadvantageous to an animal functioning at low light levels, because spatial resolution is already likely to be low. It is not surprising that tapeta are absent from the eyes of most birds, and even absent from owls because of the need to retain some degree of spatial resolution at night. That tapeta have been reported in some nightjars suggests that in these species evolution has favoured sensitivity over resolution, and that nightjars need only gross spatial detail for their night-time activities.

When not blindly trawling, nightjars normally approach individual large flying insect prey, such as moths, from below. Only occasionally are nightjars seen to swoop down on flying insects. This could be because they are better able to detect insects in silhouette against the sky. As discussed in Box 9.1, the sky is always bright enough to be visible, even on the darkest night, and thus objects can be seen in silhouette, although resolution will be low.

Together, these features provide a plausible explanation for nightjars' foraging abilities. However, these features cannot account for the social behaviour of the birds. All nightjar species are highly vocal, and it seems that social contact is maintained through vocalisation and hearing, although nothing is known of auditory sensitivity or the ability to locate and range sound sources. The plumages of nightjars often have conspicuous and relatively large white patches, and these are presumably important signals in social interactions, providing cues at only low spatial resolution.

Frogmouths and potoos

Frogmouths and potoos forage primarily beneath woodland canopies for insects. All of the species live in the tropical and subtropical climate zone. Frogmouths are distributed around Southeast Asia, Indonesia and Australia, while potoos are found in the American tropics. All species forage primarily for large insects and other invertebrates, although the larger species of frogmouths, for example Tawny Frogmouths, also take small vertebrates such as frogs and mice.

A striking feature of all frogmouths is their conspicuous and large bill which can be opened to a large gape (Figure 10.4). In most species the mouth is surrounded by prominent rictal bristles. Unlike the nightjars, the large gapes of these birds are not used for trawling insects from the air. Prey is taken primarily from the ground after pouncing from a low perch, in a similar manner to the woodland owls, but while the owls primarily take prey in their feet, frogmouths strike or capture prey directly with their large bills. The potoos do, however, take prey from the air, but rather than trawl them, these are taken 'flycatcher fashion' by sorties from a fixed perch to which the bird returns. Potoos generally live in more open woodlands than frogmouths.

Little is known about the sensory capacities of these birds. The eyes are relatively large, suggesting that vision is of importance to them. The role of hearing as a means of locating prey, especially on the ground, may be particularly important

FIGURE 10.4 A Tawny Frogmouth *Podargus strigoides* is depicted pouncing on a lizard. The large gape of frogmouths is used to scoop up such prey directly into the mouth. They usually pounce from the ground or from a low perch and take a wide range of invertebrates and vertebrates. Tawny Frogmouths are widely distributed in Australia. (Drawing by John Busby © the Artist's Estate. All Rights Reserved 2019 / Bridgeman Art Library.)

in the prey capture techniques of frogmouths. This is because any forest-floor prey must reveal its presence by the sounds it makes when moving through litter. However, nothing is known about hearing in these birds, especially their ability of locate and range sounds.

Unfortunately, only a rather sketchy picture emerges of the possible sensory ecology of these species. It is worth noting that their prey capture techniques do not seem to involve extensive flight. If flight is involved it is of short range from a fixed perch down to prey and back again, similar to a woodland owl. Clearly there may be some parallels between the sensory bases of their prey capture technique and general mobility, and that described for the woodland owls. It is worth noting that all of these species seem to be solitary and sedentary. As argued for the woodland owls, this sedentary habit may be a key behavioural adaptation that allows them to build up familiar landscapes when nocturnal light levels are high, and this knowledge can be used to successfully interpret cues at a lower spatial resolution when light levels are lowest.

This is different to the situation in nightjars. Many nightjar species breed in seasonal habitats in which suitable prey occurs in high abundance but alters markedly during the annual cycle. In consequence many nightjar species are migratory and would therefore be unable to build detailed knowledge of particular locations. However, because they exploit relatively open habitats, they do not need such detailed knowledge of the spatial structure of a particular place.

Occasional nocturnal activity

Activity at night occurs in a wide range of bird species which are usually considered diurnal. Some species sing at night to declare their presence to potential mates and rivals, some species undertake migratory journeys at night, some seabirds may come ashore and enter nest burrows at night, and there are a number of species that occasionally forage at night.

Night singing, night migration, and visiting nests at night are unlikely to be highly challenging from a sensory perspective if we accept that these activities can be achieved using only partial information, or at least that they do not require the detection of high spatial detail. After all, when engaged in these activities the birds may be more or less stationary, or flight is in open airspace, well clear of obstacles. There are, however, interesting questions about how night-migrating birds may actually guide themselves, and about how seabirds can locate their burrows when coming ashore at night. Occasional nocturnal foraging would seem to be a challenging task, but there is some good evidence on how this might be achieved.

Occasional nocturnal foraging

Two groups of birds are noted for their occasional nocturnal foraging: shorebirds (Charadriiformes), particularly species among the sandpipers and snipes (Scolopacidae), and waterfowl (Anseriformes). Many species in these taxa can be found foraging during both day and night. For some species the pattern of feeding is determined primarily by the state of the tide at coastal and estuarine foraging sites.

There is evidence that night feeding is less efficient than daytime feeding in the same species, suggesting that it is a non-preferred feeding option. In other words, the birds can do it, but they have to spend more effort, and so it may occur only when food requirements are high, or daytime feeding has not been possible. This has been demonstrated clearly in Eurasian Oystercatchers *Haematopus ostralegus*, Bar-tailed Godwits *Limosa lapponica*, Grey Plovers *Pluvialis squatarola*, and Common Redshanks *Tringa totanus*. At one study site it was shown that the feeding rate at night of Oystercatchers was on average about half that recorded on the same foraging area the following day.

Among waterfowl species, patterns of occasional night feeding have been noted. For example, among swans and geese feeding usually occurs by day, with night feeding rarely recorded. However, Brant Geese *Branta bernicla* may feed regularly at night in tidal areas but do so primarily on moonlit nights. Egyptian Geese *Alopochen aegyptiaca* appear to feed predominantly at night in freshwater habitats, while Common Shelducks *Tadorna tadorna*, which tend to be coastal, feed at night according to the tides. All dabbling duck species may feed at night, especially if they exploit tidal locations, but birds which feed in non-tidal locations feed at night primarily if they are persecuted or disturbed by people or natural predators.

The general conclusion is that among ducks and geese feeding at night tends to be a flexible response to local situations (tides, daytime disturbance, winter food shortages), rather than a preferred strategy. It may be best to consider that in these birds, and also in the shorebirds, there is the option to forage at night, but on the whole it is less efficient than daytime foraging.

How is this flexibility possible? What information do these birds have available to guide their nocturnal foraging? The answer seems to be that tactile and taste cues can be used for the location and identification of food items. But there is also some intriguing evidence that auditory cues may also play a role in the foraging of some shorebirds. In essence, all of these birds have the ability to use non-visual information for their foraging and this frees them to forage, albeit with lowered efficiency, independently of ambient light levels. However, it should be borne in mind that the foraging of all these birds takes place in open habitats largely devoid of obstacles. This means that the birds may not need long-term familiarity with particular locations in order to be able to move relatively safely.

As described in Chapter 7, shorebirds of the genus *Calidris* can use taste cues to detect the presence of chemical information left by polychaete worms or bivalve molluscs in an apparently uniform substrate. This indicates that at least some species of shorebirds can readily determine which areas of apparently uniform mud or sand will be more profitable for foraging.

Finally, mention should be made of evidence which suggests that hearing could play a role in the foraging of at least some shorebird species, notably the plovers (Charadriidae). All species of this family seem to feed occasionally at night, but their foraging typically involves taking prey from the surface, or by probing just a couple of centimetres at most into the surface. Field observations suggest that their foraging is guided primarily by visual information. However, there are observational studies which suggest that sound may also be used. This information could be derived from the sounds made by invertebrates as they move on or below the surface. Such auditory foraging has been proposed in European Golden Plovers *Pluvialis apricaria* and Northern Lapwings *Vanellus vanellus*. The evidence is rather tenuous, however, and nothing is known of hearing or sound location in these birds. If hearing is used it would function only at close range.

Regardless of whether auditory information is available, it seems clear that tactile and taste information can play an important role in the foraging behaviour of many, if not all wildfowl and shorebirds that feed at night. It would seem that this information is sufficient to allow feeding in the absence of visual information. Because these birds forage in open habitats, spatial information from vision can be rather poor but still allow the birds to move around feeding locations.

Nocturnal migration

There is a wealth of evidence that many birds undertake at least part of their migratory journeys at night. Most nocturnal migration involves bird species which are diurnally active during the rest of their life cycle. Night migration is particularly common among passerine species but not exclusive to them.

Studies of circadian rhythms in birds (mainly passerine species held in cages) have shown that at the time of year when migration usually takes place, instead of roosting, the migrating birds begin to show high levels of activity after the end of the normal daylight period. This activity is known as *migratory restlessness*. The restless behaviour is not simply random activity but shows orientation towards a particular direction which is correlated approximately with the direction in which a bird would travel if it were free to do so. Also, the time of year and the number of nights during which a bird shows migratory restlessness correlates with the time period over which the species migrates.

How do these birds determine direction when showing migratory restlessness? It seems that detecting the geomagnetic field (Chapter 8) is at least part of the explanation. It was first demonstrated more than 50 years ago that the migratory orientations of restless Eurasian Blackcaps *Sylvia atricapilla* are influenced by the orientation of the magnetic field. But it does not appear to be the sole compass cue used by the birds to determine direction on their actual journeys. There is evidence, for example, that these birds use magnetic information to calibrate another compass based on star patterns that are available during the night.

While nocturnal migrants may be using star patterns and geomagnetic information for at least their initial orientation, they are probably not flying oblivious of visual information from below them. Thus, under most night-time conditions birds seem to be able to maintain their orientation even when flying below a cloud ceiling, but they do become disoriented if they fly within cloud or fog. This suggests that some kinds of visual cues, derived either from the stars and/or moon above them or from the land/sea below, are necessary to maintain orientation. They are not just flying by compass. Furthermore, migrating birds are attracted, often fatally, to isolated illuminated structures under low-visibility conditions, suggesting strongly that at night these birds can be dominated by visual cues from below them and that they can become confused.

The phenomenon of attraction to illuminated structures would seem to suggest that these birds can be subject to considerable perceptual confusion at night if flying in the lower airspace. Thousands, if not millions, of night-migrant birds are thought to perish as a result of flying into illuminated skyscrapers in the major cities of the American northeast. However, night-migrating birds are usually flying at considerable altitude, often thousands of metres above land and water, well away from obstacles, and are thus not being called upon to conduct fine spatial discriminations. They may, in effect, be flying almost blind, and it is only under adverse weather conditions that birds fly lower – and then this near blind flight poses problems which can lead to confusion and ultimately to collisions with illuminated structures that protrude into the airspace.

Finding nests at night

Prominent among birds that visit their nests only at night are seabirds from the Procellariiformes: shearwaters, petrels, and storm petrels. These species breed colonially, nesting underground in burrows or natural rock crevices. Entering and leaving these sites only at night is regarded as a strategy to avoid aerial predators such as gulls and skuas that are active in daylight. These seabirds are vulnerable to such attacks, because their movements on land are slow and cumbersome. Some of these birds are reported as coming ashore to the nest site only under the darkest conditions when bright moonlight is absent.

How do these birds find their personal nest sites, especially in a dense colony? The sense of smell seems to play an important role, but definitive answers have been difficult to come by. This is because no one experiment has been able to control all of the possible cues in a systematic way. Olfaction, audition, and vision could all be involved, possibly used in a redundant way to ensure that the birds can find their nest no matter what conditions are present.

Chapter 6 described how the Procellariiformes have a well-developed sense of smell, and it has been demonstrated that they can use olfaction to locate their own nest burrows within a relatively crowded colony, perhaps as the primary means of doing so. However, some studies have led to the conclusion that visual cues are employed. These experiments introduced prominent visual markers near nest entrances, and they were found to influence the birds' behaviour: they could be tricked to going to the wrong burrow by moving the marker. However, these markers do not mimic the kinds of cues that would be available in natural situations. It could be that birds will learn about visual markers, but they are not what the birds primarily use under natural conditions.

Hearing could also play a role in nest burrow location through the recognition of the call of the mate inside the burrow. That such a cue is available has been shown in Manx Shearwaters *Puffinus puffinus*, with demonstrations that individuals can recognise each other's calls. However, this cue is not essential for burrow location since birds can find the correct burrow when it is occupied only by a silent chick.

Thus it seems that procellariiform seabirds, as in the other scenarios discussed in this chapter, have available to them a range of information from different senses: vision, hearing, and olfaction. Each sense is supplying only partial or spatially gross information, which of itself is not sufficient for burrow location. However, when used in a hierarchical or redundant manner this information is clearly sufficient to locate nest burrows under low ambient light levels.

Conclusion: birds in the dark

Chapters 9 and 10 have shown that birds do a lot in the dark. Even species that we commonly think of as diurnal may complete important parts of their life cycles at night. However, what they do is restricted to specific tasks. It is only a relatively small number of species that do everything at night.

It has been shown that the sources of information that may be used to guide nocturnal activity come from all of the senses. There are no simple 'super-sense' explanations of nocturnal activity. The senses of nocturnal birds are often close to the theoretical limit for their sensitivity, but this may still not be sufficient to meet all of the challenges of nocturnal environments.

One clear conclusion is that in all instances of nocturnal activity vision plays a limited role, mainly because spatial resolution is low. Other senses, and sometimes also quite specific behavioural adaptations, often allowing the accumulation of knowledge about specific locations, are seen to be at play. It is also clear that birds at night are not just doing what we might expect them to do during daylight hours. Even the fully nocturnal species, such as the woodland owls, are not doing what we would expect of their daytime counterpart species. Both the tasks that they do and the information that they use to guide them are different.

The problems for ourselves with night-time activities arise not just because of limitations of our senses. The main problem is that we want to do familiar daytime activities at night. Taking yourself to a wood at night will demonstrate that you can indeed detect a lot of information. However, your expectations of what can be achieved with that information will have to change if you want to become a night-person. To maximise your own nocturnal activity, a key lesson must be taken from the woodland owls. Stay put and learn how to correctly interpret the information available within a particular location. The traditional nocturnal poacher knew this: keeping to a local patch allowed him to travel freely and slip away from pursuers with relative ease.

Chapter 11

Birds underwater

Dive beneath the surface of a pool and take a look. You have entered a different world. Not only has the scene changed, but how you see it has also changed. Even in the clearest water your vision is altered by immersion.

If you have 'normal' vision, you will be used to seeing objects in clear focus, sharply defined. But beneath the surface your vision is blurred. Lines and edges that you expect to see clearly now appear fuzzy. Patterns of narrow stripes that are just detectable above the surface will disappear into a uniform grey blur. Why does immersion affect vision in this way?

In air, vertebrate eyes have two principal optical components, the lens and cornea, which together focus an image of the world on the retina (Chapter 3). The cornea's optical function arises because it is a curved surface that separates air from the liquid-filled chamber of the front section of the eye. The combination of a curved surface and a difference in refractive index between air and the liquid is all that it takes to make the cornea a lens, able to focus light and produce an image. However, diving beneath the surface means that the cornea effectively has water on both sides of its curved surface. The consequence is that its optical property simply disappears, leaving the eye with only the lens to focus the image. The lens is still working but, on its own its optical power is not enough to produce an image on the retina. In effect, the image is now focused behind the retina. The blurred image you see is caused by imperfectly focused rays which are on a path that would bring them to a sharp focus at a point behind the retina (Figure 11.1).

If you are already 'long-sighted' in air the image can be considered as always focused behind your retina. In effect your eyeball is too short. Entering water means that the image is produced even further behind the retina and so there is even greater blur to the image. On the other hand, if you are 'short-sighted' in air the sharp image lies in front of the retina, and your eyeball is effectively too long. Losing the power of the cornea when the eye is immersed in water may actually take the image closer to the retinal surface. The result is that you may experience a slight improvement in the quality of the image at the retina when opening your eyes underwater.

While you are always likely to notice that your vision has become blurred upon immersion, you might also notice other changes that occur at the same time. You may notice that the world has darkened slightly and that your field of view

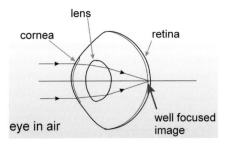

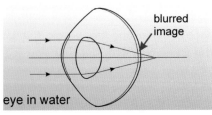

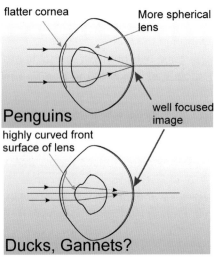

FIGURE 11.1 The problems of immersion and how it might be overcome in birds. An eye that is well focused in air (top) loses the power of the cornea upon immersion (below). An image, instead of being well focused on the retina, is in effect focused behind the eye and the image appears blurred. Two possible mechanisms may have evolved to overcome this problem in amphibious birds which move frequently between air and water. In some amphibious species the cornea is relatively flat and therefore the loss of its optical function upon immersion is less of a problem. The extra power needed to focus the image on the retina is achieved by a more spherical lens whose power needs to change little. There is good evidence for this type of mechanism in penguins and albatrosses. An alternative mechanism (bottom) is a highly malleable lens which upon immersion is pushed through a thickened iris; the front surface of the lens then bulges forward to make a highly curved and hence powerful lens. There is evidence that this occurs in some ducks, and maybe in gannets and cormorants. However, the evidence is not compelling. The bulging lens may be an artefact of the investigation procedure which has over-driven the muscles of the eye that control lens position and iris.

has narrowed. These changes are also due to the loss of the cornea as an optical structure. Losing the cornea means that the focal length of the eye increases (the focal point has moved further from the lens), but also the effective size of the pupil decreases because it is no longer magnified by the cornea. The focal length goes up and the pupil size goes down. Together these changes mean that the f-number of the eye increases and so image brightness decreases (Chapter 4). Abolishing the magnifying power of the cornea also moves the margins of the field of view towards the axis of the eye. The effect is that a smaller section of the world is projected onto the retina, and so the total visual field shrinks.

Clearly, losing the optical power of the cornea upon immersion has important effects on the vision of any eye. It presents some important challenges for all amphibious animals every time they make the transition from air to water. However, the perceptual challenges of going underwater do not stop there.

In the natural world, going through a water surface means that many conditions of the ambient light will change, and they may continue to change with increasing depth. First, natural waters are generally not crystal-clear, because they usually hold a lot of suspended particles and dissolved matter. These scatter light, resulting in reduced contrast within a scene, much in the same way that mist and fog act to reduce contrast and the penetration of light, hence reducing visibility.

Second, even a smooth water surface reflects back sunlight. Just a few centimetres below the surface light levels are reduced compared with above it. Ripples and waves reflect back even more light. Third, even clear water absorbs light. Tap water may appear transparent in a glass, but light is absorbed as it passes through water, and with increasing depth light levels decrease rapidly. It may be bright sunlight at the surface, but only a few metres below the surface light levels start to decrease noticeably because of this absorption, and they continue to decrease with increasing depth. Fourth, even pure water absorbs light differentially with respect to wavelength. Again, this is not apparent when looking at water in a glass, but water absorbs strongly at the redder end of the spectrum and is most transparent to the blues, but highly absorbing to very short (UV) wavelengths. This has the effect that with increasing depth the spectrum of ambient light narrows towards the blues. A TV programme showing life at ocean depths often depicts a colourful world, but these colours are visible only because of the artificial lights used for the filming. In nature at any reasonable depth all will be in shades of blue; bright greens, yellows, and reds are not present (Figure 11.2).

Because birds are air-breathers, the majority of them stay down for only short periods, a few minutes at most. This means that foraging is partitioned into discrete bouts and food items have to be detected and procured very rapidly, with limited time to get hold of an item once it has been detected. These short periods underwater mean that the birds must transition very rapidly between different light levels both as they dive and as they travel back to the surface. Such rapid transformation in light levels does not allow sufficient time for the process of retinal dark adaptation to occur (Chapter 3). Complete dark adaptation, resulting

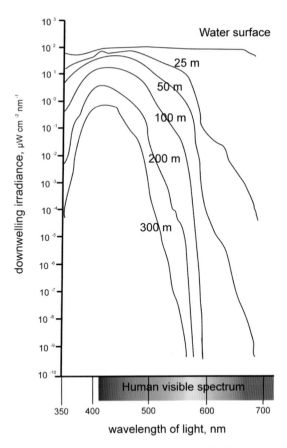

FIGURE 11.2 Sunlight falling on the surface of water is absorbed significantly as it penetrates beneath the surface. This absorption is selective of the wavelength of light. Its effect is cumulative, and ambient light characteristics change dramatically with increasing depth. These changes pose a challenge for the vision of any species which dives even to relatively shallow depths. It is especially challenging for species which are amphibious because they transition regularly and rapidly between two quite different light environments. The sunlight falling on the surface contains all light wavelengths of the visible spectrum at more or less equal intensity. However, absorption of light by water changes this dramatically. Not only do light levels fall, selective absorption by water results in a narrowing of the spectrum of ambient light which becomes progressively more blue-violet. Even at a depth of 25 m light in the yellow, orange, and red parts of the spectrum is strongly absorbed. Ultraviolet light is also absorbed. This diagram is based on measurements of the deep pure waters of Crater Lake, Oregon. At a depth of 100 m the spectrum of light is very narrow, and even at the peak of light transmission light levels have fallen 10-fold and the bulk of the light falling on the surface has been lost. Some species of auks regularly dive to 100 m and some penguins dive to 300 m. At these depths the ambient light is almost exclusively made up of deep blue and purple wavelengths and very little light penetrates, so even at midday the birds will experience light levels as low as those experienced at night at the surface. These absorption curves are for pure water; most natural water bodies contain large amounts of microorganisms and dissolved material which further alter the spectrum and lead to the absorption of much more light, so that night-time light levels may be experienced at shallower depths.

in the retina achieving highest sensitivity, takes about 40 minutes. Many diving birds stay down for less than 2 minutes, and over that time sensitivity gains are relatively small. These birds do not have the luxury of their retinas adjusting to the slow changes in ambient light levels that naturally occur through dusk and dawn. They must cope with changes in light levels that are far too fast for their retinas to track.

Clearly, going beneath a water surface poses a set of perceptual challenges that are quite different from those posed by the world above the surface. In a nutshell, visual information is rapidly lost due to a combination of factors: image defocus, light scatter, light decrease, and a narrowing of the light spectrum. So why have so many bird species evolved to conduct the vital tasks of foraging under such apparently challenging sensory conditions, and how do they cope?

It is shown in this chapter that these sensory challenges are not met with a single adaptation; there are no super-sense explanations. Rather, as with the examples of nocturnal birds, there is good evidence that the sensory challenges of underwater activity are met with a combination of behavioural strategies and sensory adaptations. However, birds underwater have even fewer sources of information available to them than nocturnal birds in terrestrial environments. It seems that only vision is used to guide the foraging of birds that take active prey at or below the surface of water. Sometimes this visual information is so restricted that some birds may rely upon random encounters with their prey, while others force their prey to reveal themselves.

Taste cues may be employed to verify a located food, and a few species may use tactile cues to help in the actual location and capture of their food. However, aquatic birds do not have available to them the sophisticated active sonar systems that are used by some aquatic mammals, or the highly sensitive tactile systems based on vibrissae (whiskers) employed by others. There is no evidence that birds have access to the electrical cues that are used by some fishes to orient themselves or locate prey in highly turbid waters. There are no demonstrations that any birds use olfaction or sound information below water. Ears that work in air probably cannot work with any degree of acoustic accuracy in water. It is true that humans can detect sounds underwater, but the information is very restricted, and the sounds are very difficult to locate. The same is likely to be true of birds, but definitive experiments have yet to be conducted.

Exploiting aquatic foods

Water bodies are widespread, covering much more of the earth's surface than land, and many water bodies are full of life, providing a very rich source of food. Because of this food abundance it is not surprising that natural selection favoured the exploitation of these resources by birds as well as other animals. However, because of the sensory and locomotory challenges of the aquatic environment, it is equally unsurprising that species have had to become highly adapted to these challenges. The consequence of this is that the majority of amphibious bird species feed exclusively on food taken from below the water surface by diving. There seem to be no species which both feed above water and also dive below it to feed. However, many aquatic foragers feed on food items at or just below a water surface, taking food through the surface, and these birds may be less specialised in their diets, relying on both terrestrial and aquatic food resources.

It is worth noting that some of these aquatic species could have been considered in the previous chapter on nocturnal behaviour. Many aquatic species occasionally forage at night, and some species habitually dive to such depths that even when foraging during daytime they actually experience night-time light levels.

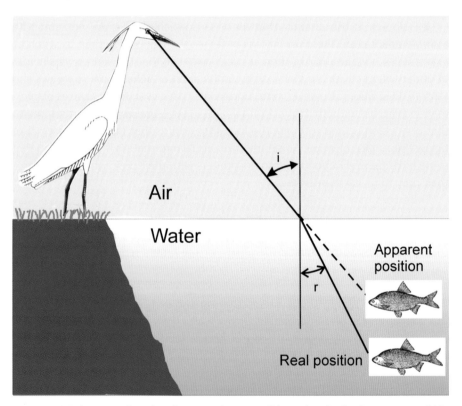

FIGURE 11.3 An egret foraging for prey detects a fish and waits for it to come within striking distance. However, refraction of light at the air–water interface means that the fish prey appears nearer the surface than it actually is. Light from the real position of the fish is refracted according to Snell's Law. The angle *i* is always greater than the angle *r*, raising the apparent position of the prey nearer to the surface. The greater the angle *i*, the greater the difference between real and apparent positions. The bird must compensate for this distortion of position, as it will usually have only one chance to catch a prey item. An attack will trigger the prey to make a rapid escape response. The situation depicted here is idealised in that the water surface is undisturbed. Any ripples will distort the apparent position of the object unpredictably, and its position will appear to change even when it is not moving.

Those birds which take prey from the water surface or just below it may make shallow dives to secure the prey, but many species take prey by just poking or probing their bill through the surface. The sensory challenges for these birds are different to those faced by divers. Their problems arise because of the difficulty of seeing through the water surface or detecting signs at the surface that prey lies beneath. The principal sensory challenge for these birds is produced by the water surface itself and the fact that shallow waters can be very turbid.

At the surface, items below may be obscured by the distorting effect of surface ripples and by masking reflections. Even clear still water presents a significant problem owing to the refraction of light at the surface. This produces a disparity between the actual and the apparent position of an object when viewed through the surface. These are the 'bent-stick' and 'apparent-depth' problems which fascinate children when they first come across them. These problems cannot be overcome in a direct sensory way. We must learn to overcome the distortions, and this learning needs a lot of practice, since the apparent position of a target varies both with its absolute depth and with the angle at which it is viewed. These same problems are faced by birds such as herons when they attempt to catch a frog through the surface of a pond, or a kingfisher preparing to dive on a fish (Figure 11.3).

There are a few bird species which have evolved to take prey items from surface waters, bypassing these light distortions and using non-visual information for prey detection. Prime among such birds are the three species of skimmers (*Rynchops*) which use tactile cues to register the presence of an item below the surface. Other birds, notably some plovers, bypass the problems of water surfaces by employing a foraging technique called foot-trembling. This can force prey lurking in highly turbid water, or in near-liquid mud and sand, to reveal itself and be caught by visually guided pecking.

Who forages for what?

Subdividing aquatic foragers is a difficult task. The number of species is large, running into hundreds, and they occur across a number of families and orders. Some species have rather flexible foraging techniques and vary their diet, depending on specific situations exploited at different times of the year. However, they can all probably be placed within one of five main groups, according to the type of prey and the sensory problems that they face in pursuing them (Figure 11.4).

One group of species dive from the water surface to find sessile prey, usually molluscs. Such prey is not evasive, but it must be detected and then prised from a solid surface. The birds that feed in this way are ducks (Anatidae). They include scaup and pochard species *Aythya* spp., Long-tailed Duck *Clangula hyemalis*, Harlequin Duck *Histrionicus histrionicus*, and eider species *Somateria* spp.

A very different type of foraging challenge is faced by bird species which dive from the surface in pursuit of evasive fish prey in deep and relatively clear oceanic

FIGURE 11.4 Birds that take prey underwater exploit different types of prey which are taken from different types of positions and at different depths. Five main types of prey/foraging situations have been identified. Greater Scaup (top left) are an example of species that take mainly sessile prey fixed to hard surfaces. Auks and penguins (bottom) dive from the surface to take evasive fish prey at depth in relatively clear oceanic waters. Divers (loons, top right) also take evasive prey but usually at more shallow depths in more turbid water in lakes and close to the coast. Cormorants and shags (middle left) also surface-dive and mainly take prey that is hidden in the bottom substrate or lurking among rocks and roots, often in highly turbid waters. Kingfishers (middle right) are examples of species which make relatively shallow plunge-dives into clearer waters; they typically detect a prey item from above the surface and take it either on entering the water or when it is seen in silhouette from below as the bird regains the surface. (Photo of Greater Scaup *Aythya marila* by Calibas [CC BY-SA 4.0], Great Northern Diver (Common Loon) *Gavia immer* by John Picken [CC BY 2.0], Great Cormorant *Phalacrocorax carbo* by Tim Evanson [CC BY-SA 2.0], Giant Kingfisher *Megaceryle maxima* by Charles J. Sharp [CC BY-SA 4.0], Humboldt Penguin *Spheniscus humboldti* by Wilfried Wittkowsky [CC BY-SA 3.0], Common Guillemot (Common Murre) *Uria aalge* by Ómar Runólfsson [CC BY-SA 2.0].)

waters. These are taken from the water column at a range of depths down to 300 m, with many of the species typically diving to between 50 and 100 m. These include the penguins (Spheniscidae) and auks (Alcidae).

A slightly different task is faced by species which dive from the surface to prey upon evasive fish in predominantly shallow depths, usually less than 10 m. This occurs typically in coastal waters and sometimes in estuaries. Prey is usually taken from the water column but these waters, because of their proximity to land, can be turbid or highly coloured due to dissolved matter. The most prominent group that feed in this way are the divers (also known as loons, Gaviidae), but they also take prey in the clear but often highly coloured waters of small lakes and lochs.

An intriguing group of birds dive to take evasive fish prey at shallow to mid-depths. These do not rely on seeing prey at any distance, rather they may disturb it from the bottom or from hiding places, such as among tree roots or rocks. This group includes the more than 40 species of cormorants, shags, and darters (Phalacrocoracidae, Anhingidae) and the 20 species of grebes (Podicipedidae). This foraging may take place both in relatively clear and in turbid coastal and inland waters.

The final group of species are those which are noted for capturing prey following spectacular plunge-dives which penetrate to relatively shallow depths. They use this technique to take either evasive prey such as fish or slow-moving prey. However, individuals sometimes also pick prey from the surface. Among these species are the eight species of pelicans (Pelecanidae), the ten species of gannets and boobies (Sulidae), and the kingfishers (Alcedinidae). These have become well known through spectacular filming of their plunge-diving. However, this group also includes the many species of terns in the family Laridae, albatrosses (Diomedeidae), petrels and shearwaters (Procellariidae), northern storm petrels (Hydrobatidae), the three species of tropicbirds which make up the order Phaethontiformes, and the four species of diving petrels in the genus *Pelecanoides*. All of these species may make shallow plunge-dives or pluck prey such as squid from the surface.

This summary shows that foraging for aquatic prey occurs in a wide range of bird species from different orders and families. It also involves different prey types and different foraging techniques. Foraging in each of these different ways is likely to pose different sensory challenges for the detection and capture of food items. Thus, underwater foraging is far from a uniform task. Some foraging would seem to be more exacting in that it involves the detection of mobile prey which may have evasive strategies to avoid capture. Some foraging poses less exacting sensory challenges in that it involves sessile foods. Although detection of sessile foods may be challenging, once found it does not have to be pursued and caught.

The foraging of the last group listed above, the plunge-divers and surface-grabbers, differs from all others in that prey is often, perhaps always, detected from above the water surface, as the birds fly over, sit on a perch above the water, or sit on the water surface. In all other types of aquatic foraging the search for food does not begin until the birds are underwater. Also, for this last group, prey items may be detected either as individual items or as concentrations. Indeed, as described in Chapter 6, among the petrels, shearwaters, and storm petrels (Procellariidae, Hydrobatidae) olfactory cues can play a key role in the detection of profitable feeding areas in the open ocean where food is highly concentrated.

Foraging underwater

No matter how well focused the vision of any animal underwater, the nature of the environment below the surface strongly suggests that vision cannot be reliably used for the detection of objects at a distance. The lower ambient light levels, the absorption of light, the turbidity produced by suspended particles, all mitigate against aquatic birds ever being able to behave like predatory birds in air. Detecting food items at a distance, and being able to pursue them at speed, is a strategy only for birds of the air. As described below, underwater birds must use other cues and other techniques to find and secure their foods.

Ducks

Tactile cues are an obvious source of information that can supplement vision, especially under low light levels or in turbid conditions. Indeed, it has already been shown that tactile cues play a key role in the foraging of ducks and shorebirds at night when visual information is lacking or restricted (Chapters 9 and 10). However, the tactile senses are not telereceptive: they cannot provide information about objects remote from the body. No matter how compromised vision may be, it is the only sense that can provide a bird underwater with information, albeit sparse information, about objects at any distance from it. These limitations are exemplified by the underwater foraging of some duck species.

Many ducks have a bill-tip organ that is capable of making fine tactile discriminations of objects at the bill tip (Chapter 7). That tactile information is used to detect specific prey items underwater is exemplified by investigations of foraging in Blue Ducks, an endemic and endangered species of New Zealand. It seems likely that these ducks use tactile cues when trying to retrieve prey items from the surface of rocks or hidden between stones and boulders, but they are also able to switch to vision to catch mobile prey in the water column. Blue Ducks indicate what may be possible with other duck species, such as the scaups, eiders, and Long-tailed Ducks.

Blue Ducks inhabit forested headwater catchments of rivers with medium to steep gradients. These waters are rich in invertebrates which depend upon unpolluted and well oxygenated water. In such waters Blue Ducks feed on a wide variety of aquatic invertebrates and occasionally on small amounts of plant material. The bird's scientific species name *malacorhynchos* means 'soft-bill', referring to the presence of a specialised structure either side of the tip of the maxilla (Figure 11.5). This consists of a pair of flexible flaps of soft keratin that overhang the mandible when the bill is closed. The flaps contain Herbst's corpuscles (Chapter 7), and

FIGURE 11.5 Flexible flaps of soft keratin overhang the mandible of Blue Ducks *Hymenolaimus malacorhynchos*. They contain Herbst's corpuscles and the flaps probably function as specialised tactile receptors that aid in the detection of sessile prey attached to rocks. (Photos by Bernard Spragg [public domain] and the author.)

these tactile receptors can function to detect when the flaps are deformed as they touch or move over a surface. They probably serve to extend the region of tactile sensitivity from just around the bill tip to along the sides of the bill.

Studies of Blue Ducks' foraging behaviours and diet demonstrate that foraging occurs within specific underwater microhabitats and that the birds adopt different foraging strategies to exploit different types of prey.

The first prey type are invertebrates attached to rock surfaces. These are unevenly spaced, suggesting that the birds cannot forage on them simply by grazing over rock surfaces, but instead have to detect individual items. Tactile information from the bill is likely to be the primary cue used to guide this behaviour. Hence Blue Ducks can 'forage blindly' on, between, and beneath rocks for these prey types, exploring with their bill into crevices that they cannot see into.

The second prey type are larger and more nutritious items: free-swimming mayfly and stonefly larvae. The ducks forage on these at a particular time of the year. Taking these, even at close range, is a visually demanding task, for the prey must be detected and grasped in the bill with precision. It has been shown that Blue Ducks have more frontally placed eyes than ducks that feed exclusively using the sense of touch (for example, Mallards, Shovelers and Pink-eared Ducks). As described in Chapter 5, these exclusively tactile feeders have eyes positioned well on top of the head and are unable to use vision for accurate visual guidance of their bills. Blue Ducks, however, can see their own bill tips and use vision to guide their bills at very close range to take the small individual items from the water column. Thus, Blue Ducks are able to switch between prey types because they have available to them both tactile cues and a visual field that allows accurate target detection at close range. Depending on prey availability at different times of the year, Blue Ducks can switch from sessile prey hidden in crevices, or obscured in turbid waters, to the more visually demanding task of capturing items moving freely in the water column.

Cormorants

The fish that Great Cormorants prey upon are often hidden, lurking among tree roots and rocks or lying flat on the substrate. This prey is often highly cryptic or of low contrast with the background, especially when the birds are seeking fish in the turbid waters of estuaries, rivers, and lakes. Cormorants do not forage at great depths, usually just a few metres, and the duration of their dives is relatively short, 30–40 seconds. Some populations at high latitudes in Greenland have been recorded as foraging at night in the middle of winter when days are short. Preying upon hidden, cryptic prey, and sometimes at night, would seem to present particularly testing sensory challenges. However, Great Cormorants are credited with being the most efficient aquatic marine foragers of fish. This has been quantified as grams of fish captured per minute underwater. Foraging efficiency can also be high when birds forage in highly turbid waters. Cormorants usually forage alone, although in some locations they may forage cooperatively when they force aggregations of fish towards the surface and attack them from below.

The most detailed information on vision underwater in any bird comes from studies of visual acuity and the visual fields of Great Cormorants. This work produced some surprising results. In particular, it showed that acuity below water is much lower than might be expected for an efficient visually guided forager. Acuity was measured using a training technique in which birds chose between high-contrast grating patterns (Chapter 2). Measured in this way, the acuity of cormorants is about 45 times lower than that of an eagle, and 5 times lower than a Rock Dove. The acuity of cormorants underwater is in fact very similar to the highest acuity achieved underwater by young humans without a face mask. This indicates that even when cormorants are foraging in brightly lit clear waters they do not see fine detail, so that only relatively large shapes will be detectable at a distance of a few metres.

It has been possible to use this acuity data and combine it with data on how the resolution of cormorants is affected by target contrast, to come up with 'a cormorant's-eye view' of a typical prey item (Figure 11.6). These modelled cormorant-eye views include prey at different distances and with different degrees of contrast. They show that only at distances of less than 1 m can a fish, of the size typically taken by cormorants, be seen in any detail. Beyond this distance a prey fish will appear no more than a faint blur. Furthermore, most fish prey is cryptically coloured and will have very low contrast against the background. In some parts of their range, cormorants feed upon sculpins (also known as bullheads or sea scorpions). These fish are renowned as examples of species which have highly cryptic body shapes and skin patterns. They are often very difficult for humans to detect, even at very close range.

So how can cormorants be so efficient in the detection and capture of fish? They seem to be far from well equipped visually for the task. They cannot be using a predation technique of the kind used by terrestrial raptors: they cannot detect prey at a distance using high-resolution vision and then hunt it down. As in the examples of nocturnal birds, the answer seems to lie in a combination of rather minimal sensory information and a specific foraging technique.

Viewing distance

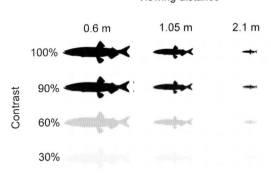

FIGURE 11.6 Cormorant-eye views. A simulation of how a fish model might appear to a cormorant at different viewing distances and at different degrees of contrast with the background. The model fish is based on a 10 cm long Capelin *Mallotus villosus*, a species commonly taken by Great Cormorants *Phalacrocorax carbo*. The acuity of the cormorant is based on behavioural measures and assumes the highest acuity of this species underwater. The water is clear and the light level is 10 lux, which is at the lower end of the daylight range. The simulation shows that apart from the high-contrast fish viewed at close range, all target fish will appear quite indistinct. This gives rise to the idea that cormorants will frequently be lunging at an escaping blur when foraging, especially when water is turbid, rather than a distinct fish shape.

The technique that cormorants use is described as 'flush-foraging close-capture'. Using this technique, they force hidden fish to make an escape response; they flush it out. Prey may be hidden among rocks or tree roots or sheltering on or within different substrate types. This provoking of a fish into making an escape response is achieved by the bird poking with its bill into nooks and crannies, or at the substrate, with darting movements using its long flexible neck. This technique has been directly observed in foraging European Shags *Phalacrocorax aristotelis*, with film showing birds poking their bills into sandy substrates in order to force fish hiding there to make an escape. In this way prey is forced to escape and is taken immediately that it does so. Capture is mainly achieved by rapid neck-extension in a fashion similar to the rapid neck-extension prey capture technique used by herons. Typically, herons have only one opportunity to capture an individual prey animal or it will escape, possibly for good. Such single-opportunity capture is also likely to be the case in cormorants.

Cormorants can travel rapidly underwater, but they are also highly manoeuvrable. When investigating objects, they can turn in tight circles, usually holding the neck retracted. Swimming in pursuit of an escaping fish may not be possible, as it may take too long to get going. However, rapid neck-extension is probably sufficient for prey capture in many situations. Given the low light levels, the turbid conditions of some water bodies, the cryptic nature of the prey, and the low acuity of the birds, it seems remarkable that anything can be caught, but the technique is clearly efficient.

There is, however, one problem with this technique. The bird may not be able to identify the prey item before it is caught. There is simply not enough time or information for that. Cormorants may often be simply lunging at an 'escaping blur'. Of course, if the blur is escaping it is highly likely to be edible food. Much in the same way that a leaf-litter rustle is highly likely to indicate food for a woodland owl, the bird does not know what it might be catching until the prey has been seized.

Much to the annoyance of anglers watching from the shore, cormorants typically bring their captured prey to the surface, and appear to hold it aloft in the bill for the anglers to admire. This may be important to allow the cormorant to identify the 'escaping blur' that it has just caught, as well as necessary for placing it into position for swallowing. The visual fields of cormorants facilitate this by allowing the birds to see what lies between their open mandibles (Figure 11.7).

This foraging technique of cormorants does not preclude them from taking prey that are detected in clear water out in the open; they may well do this if opportunities arise. However, it does allow them to capture prey when none can be seen initially. It may well be responsible for the extensive distribution of Great Cormorants across the world, within which they occupy a very wide range of fresh and marine water types and habitats. Wherever cormorants find themselves, they can simply start to forage, no matter how turbid the water, and test for the availability of hidden prey. Indeed, this technique may well explain how cormorants can

FIGURE 11.7 A Great Cormorant looks between its open mandibles. The main function of this ability is probably to allow the examination of prey when it is brought to the surface prior to swallowing. The bird may have caught an 'escaping blur' (see Figure 11.6) and will need to verify the identity of the prey item as well as positioning it for swallowing.

take the highly cryptic sculpins. They simply poke about until a fish is disturbed; there is no need for a cormorant to break the camouflage with its vision.

The habit of cooperative foraging reported in some cormorant populations may have the function of concentrating fish close to the surface where they can be seen, possibly en masse, in silhouette against down-welling light. Individual fish do not have to be detected, in fact the fish are likely to appear to cormorants as a large swirling blurred mass, but individual fish can be caught by rapid neck-driven lunges into the aggregation of fish.

This flush-foraging close-capture technique, rather than the pursuit of individual fish through the water column, could well be one employed by other birds that forage underwater. These might include the divers and grebes, since they typically forage at shallow depths where they can disturb prey hiding or resting on substrates, and they too are able to take prey in highly coloured and turbid waters. There is evidence that Great Crested Grebes *Podiceps cristatus* may also disturb fish and drive them towards the surface where they can be detected in silhouette from below.

Penguins

The flush-foraging close-capture technique used by cormorants cannot apply to the foraging of birds that take prey from the water column. Flush-foraging works only if fish and other prey have somewhere to hide or settle. Prey in the water column has to be detected visually at a distance or encountered at random. This is the kind of foraging task that penguins face. They forage in the open water, sometimes at considerable depth. Even when tied to land by the demands of nesting they forage in the open seas, not within the more turbid shallower waters near the shore where they are more prone to predator attacks. Outside of the breeding and moulting periods penguins live continuously in the open ocean.

Evidence of the role of vision in the foraging of penguins comes mainly from investigations of the eye structure and visual fields in Humboldt Penguins *Spheniscus humboldti* and King Penguins *Aptenodytes patagonicus*. Unfortunately, there are no measures of visual acuity in any penguin species.

The eyes of Humboldt and King Penguins show a key structural feature which suggests that their optical design is shaped primarily by the aquatic environment. In these eyes there is a marked imbalance between the optical powers of the lens and the cornea. In terrestrial bird species, including owls, starlings, and ostriches, the cornea and the lens contribute approximately equally to the overall power of the eye (Chapter 4). This is also the case in our own eyes. It is for this reason that losing the power of our cornea upon immersion changes vision noticeably, for the lens by itself cannot bring the image onto the retina. In penguins, however, the balance between the lens and cornea is very different, with the power of the cornea only one-third that of the lens.

The importance of this imbalance is that when a penguin dives, and so loses the power of its cornea, this has a much smaller effect on the optical performance of the eye as a whole (see Figure 11.1). If the eye is well focused in air there will still be some defocus as it goes below the surface, but the effect will be far less than in our own eyes. Furthermore, a detailed model of the Humboldt Penguin eye's optical system indicates that their eyes are slightly short-sighted (myopic) in air but well focused (emmetropic) in water. Losing the power of the cornea upon immersion actually improves the focus of the eye.

A particularly intriguing clue to the role of vision in the foraging of penguins comes from the observation of the unusual iris in King Penguins. The iris is the structure that controls the size of the pupil. In many birds it is highly coloured and makes the eyes conspicuous, but in King Penguins the iris is dark brown/black and so cannot be readily seen. Close inspection, however, shows that the King Penguin's iris is unusual in that it can be stopped down so much that the pupil becomes a pinhole. However, the pupil can also be dilated to almost the full diameter of the cornea, providing a relatively huge pupil. As it dilates from a pinhole the pupil first becomes square, then hexagonal, and finally circular. The pupils of other penguin species are often seen to be very small, almost pinhole size, but they have not been studied in detail.

The size range of the King Penguin's pupil is exceptional, and it has important consequences for vision. It can alter the brightness of the retinal image over a range of about 300-fold. This is a huge range; by comparison, image brightness due to pupil diameter change in the eyes of Rock Doves and humans varies over a range of only 16-fold. Photographers who are familiar with thinking about the performance of their lenses in terms of f-numbers should note that the King Penguin has an eye whose f-number can change between 1.8 and 31, a very wide range found only in specialised photographic lenses.

This huge range in pupil size in King Penguins may serve two functions. First, by being able to close down to such a small diameter, the eye can become a pinhole camera, bypassing the optics and producing a well-focused retinal image in air, thus overcoming any myopia. Secondly, and perhaps more importantly, the closed-down pinhole aperture produces a very dim retinal image that can maintain the retina in a dark-adapted state. Pinhole pupil apertures have long been recognised as having such a role in the eyes of nocturnal mammals and reptiles. These animals typically have a well-stopped-down pupil when exposed to the sun. For example, it is almost impossible to see a cat's pupil when it is basking in the sun.

King Penguins forage by both night and day. At night they forage at relatively shallow depths (maximum of 30–40 m) but during the day they dive to greater depths; dives of between 100 and 300 m are typically recorded. At these depths light levels are much reduced, in fact they are within the range of light levels that occur at the surface at night. So these birds are effectively always foraging at night-time, or at least twilight, light levels. They forage at different depths to follow the vertical migration of their prey, which move nearer the surface at night but stay at depth during the day. To forage at such depths requires these penguins' eyes to be adequately dark adapted at the time they reach the twilight-night-time zone.

They need to have maximum sensitivity to be able to gain as much information as possible.

The problem for King Penguins is that they must dive very rapidly in order to have maximum time at foraging depth. During the day they get to their low-light forage zones in a little over 1 minute. Consequently, there is no time for the retina to become dark adapted on the way down, the eye must be pre-adapted. Keeping the pupil to a pinhole at the surface means that the eyes can always be dark adapted. As the bird dives, the pupil can be opened up. This allows the eye to gain as much light as possible as it rapidly fades with increasing depth. On the return to the surface the pupil can be instantly closed, thus preserving the dark-adapted state of the retina ready for the next dive. Using this pupil mechanism, King Penguins can easily transition from the light of the surface to the dark of the depths and always be well adapted to the low ambient light levels where their foraging takes place.

Of course, even with an adequately dark-adapted retina, spatial resolution is likely to be low at these twilight-night-time light levels. The foraging success of King Penguins at these depths probably also depends on the specific type of prey that is taken. Myctophids (lantern fish) are the principal diet of King Penguins. These fish are small (46–79 mm in length), but they have photophores on their body surfaces. The fish are patchily distributed in shoals. To detect them it seems likely that King Penguins do not rely on detecting fish, but rather on detecting the photophores which indicate the presence of fish. Photophores will show essentially as point sources of light against a dark background. For the detection of point light sources, it is the absolute size of the pupil aperture, rather than the f-number of the eyes, which determines sensitivity. This may be another reason why King Penguins' eyes have corneas and pupils of large diameter.

These small myctophid fish are widely dispersed in the water column, yet they are ingested at a high rate by King Penguins. It seems likely, therefore, that the penguins must search for their prey within a wide sector of the space surrounding them. King Penguins have a broad visual field that correlates well with this requirement. The foraging task of King Penguins can perhaps best be characterised as requiring the location of pinpoints of light at any position about them. Such foraging is accomplished through using large eyes whose retinas are maintained permanently dark adapted throughout the many dives of a foraging session. Even though the birds return to a brightly lit surface between dives they are always well prepared for the darkness below.

Auks

Like penguins, auks (Alcidae) take fish from the water column. They are sometimes considered the northern-hemisphere ecological equivalents of the southern hemisphere's penguins. Some species, notably Common Guillemots (Common Murres), are known to forage at night and descend to depths of up to 150 m. Therefore, auks face visual challenges similar to those described for penguins: foraging at low light levels with a rapid transition from bright sunlight to twilight

light levels within a dive. Unfortunately, nothing is known about the eye structure of auks and there have not been any investigations of their underwater vision. However, there is some information on visual fields in two auk species, Common Guillemot and Atlantic Puffin (Figure 11.8). There are also some intriguing observations of foraging by Guillemots at night. Together these give some clues about the sensory ecology of foraging in auks.

FIGURE 11.8 An Atlantic Puffin *Fratercula arctica* and a Common Guillemot (Common Murre) *Uria aalge*, two species of auks that are often associated when using the same locations for breeding. However, they forage in different ways and on different prey. Apart from the obvious difference in their bills a particularly notable feature of these photographs is their eyes. Most notably they differ in their positions in the skull and in the directions in which they project. This is reflected in significant differences in their visual fields (see Figure 11.9), which can be related to differences in their foraging ecology.

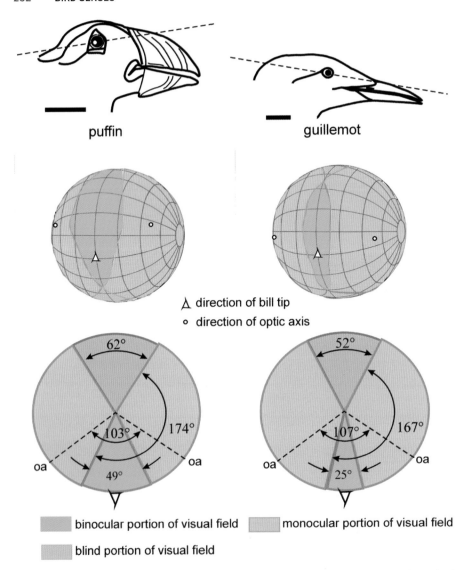

FIGURE 11.9 The visual fields of Puffins and Guillemots show both subtle and marked differences which can be related to differences in the foraging ecology of these two species. In the upper diagrams the dashed line indicates the plane of the section through the visual fields which are shown in the bottom pair of diagrams. Most notable is that while the width of the field of each eye is similar in Puffins and Guillemots, the way that the eyes are placed in the skull results in binocular fields of significantly different width. The position and extent of the binocular fields are shown in the middle two diagrams, indicating that Puffins have the widest part of their binocular field projecting upwards and forwards, while in Guillemots it projects directly forwards.

The visual fields of Guillemots and Puffins differ significantly in ways that can be related to their diets (Figure 11.9). Guillemots feed on small shoaling fish throughout the year. They have a visual field in which binocular vision is centred around the bill. This is similar to the configuration of visual fields in terrestrial species which use vision to control bill position during pecking or when lunging at prey. At the very least this suggests that Guillemots use vision to detect and take fish from the water column, at least when there is sufficient light to guide them.

Puffins, on the other hand, are less specialised in their diet, and they probably switch between prey types during their annual cycle. Thus, the well-known images of Puffins holding multiple sand eels in their bills are associated with the provisioning of young, but they do not tell us what the birds eat when away from the breeding colonies. Outside of the breeding season Puffins go to the open ocean and are probably not highly dependent upon sand eels. At this time, it seems that they switch their diet and take much smaller prey items including crustacea, polychaetes (bristle worms), chaetognathids (arrow worms), and squid. Thus, invertebrates may make up a major component of a Puffin's winter diet.

The perceptual tasks involved in locating and catching such prey appear challenging, since these planktonic animals are mobile, semi-transparent, and distributed in the water column. It seems that the visual challenges of this exacting task can be related to the broader binocular fields found in Puffins, and to the ways in which the field projects more upwards rather than being centred around the bill, as in Guillemots. This more upward-looking vision will facilitate small prey items being seen in silhouette against the down-welling light. Once a prey item is detected it can be grasped in the bill by swinging the head upwards.

Although Puffins appear to have large bills they are laterally flattened, the gape width is not large. Therefore, trawling for small prey is an unlikely strategy to be used by Puffins, and in any case trawling may not be possible without a filter mechanism of some kind. For Puffins, individual items, although small, require detection and selective seizure, although sometimes these invertebrates can occur in concentrations. The perceptual challenges of taking the larger fish prey items which predominate in the breeding season can be met within the parameters set by the challenges of taking these smaller prey, but not vice-versa.

There is intriguing evidence that Common Guillemots may be able to forage blind, relying upon random encounters with fish at depth. This has been argued from data on the ingestion rates and energy expenditure of Guillemots. It seems that individual fish are not sought but just encountered randomly, at least at night. This hunting technique is only viable if the fish prey occurs at a sufficient density in the water column. However, the threshold density for successful random-encounter foraging may in fact be relatively low and does not require dense concentrations. How a Guillemot registers a random encounter with a fish could be through vision at very close range, and/or through simple tactile cues as a fish is struck, most likely by the bill.

Clearly, such random searching, and virtually blind encounters with prey, could be a technique also employed by penguins. In fact, it could be used by any bird species which take fish from the open water column, as long as the prey occurs at a sufficiently high density. The technique does not, of course, preclude birds using vision to detect individual items, but it does decouple prey capture from light level.

Foraging at the surface

A large number of bird species forage for prey which is at, or just below, the surface of water. Other species take prey that is just below the surface in water-saturated mud and sand. In these situations, foragers always work from above the surface and so the focus of their eyes is not compromised by the loss of the cornea's refractive power. This would seem to be a less challenging way of getting access to aquatic prey. However, foraging through a water surface presents new problems. Surface reflections and ripples distort what can be seen below. Upwelling light is different from the incident light from the sky above. Refraction of light at the surface means that an object's true position is distorted. Objects appear nearer the surface than they actually are, and their apparent positions depend upon the angle of view.

Herons and kingfishers

Herons (Ardeidae) and kingfishers (Alcedinidae) are the two prominent groups that face these sensory challenges of foraging through a water surface, and there are some clues as to how they overcome them.

In herons, a key adaptation which helps in their pursuit of prey below a water surface is found in their visual fields. Herons have comprehensive vision of the hemisphere in front of them. While many birds see much of the world in front, none see as much as herons. In most birds visual coverage drops away to the sides at or below the bill. However, herons see directly below their bill. This visual coverage is similar to the comprehensive vision found in Mallards, but rather than looking around and above the head, the field in herons is tipped forward through 90° so they see everything that lies forward of them, including right down to the ground or the water beneath them.

The well-known illustrations of bitterns (*Botaurus* spp.) looking below their bill directly at the observer are a manifestation of this. We know about this because bitterns have the habit of trying to conceal themselves in reeds and stretch their neck vertically with the bill pointing skywards in an effective cryptic posture. Photographs of birds in this posture show the eyes apparently looking forward, but in fact they are looking directly below the bill. The full utility of this arrangement is that when the bird holds its head horizontal it can see all that is below it. This allows a heron to stand motionless and let prey come to it while maintaining surveillance of all that is below. Only when a prey item is within strike range does a heron lunge its bill forward to take the prey. This is important because most prey items are highly evasive and there is only one chance to strike; miss, and it will be gone.

This configuration of visual fields has been described in a number of heron species; it is certainly not unique to bitterns (see Figure 5.3). It is a key adaptation for the foraging success of herons, allowing prey to come to the heron, or to be approached very stealthily, and to be caught in a single strike. The head is propelled forward very rapidly, aided by a special joint in the neck. The lunge of the bill must be right on target and timing must be accurate. To achieve this, herons must have relatively clear water to forage in. But crucially they must also compensate for the effect of refraction when looking through the surface. They must strike at the actual position of the prey, not its apparent position (see Figure 11.3).

That herons can do this must be true. It has been investigated experimentally in Little Egrets *Egretta garzetta* and Squacco Herons *Ardeola ralloides*. These studies showed that herons are correcting for refraction in a flexible way, indicated by the fact that they are able to strike successfully from various angles.

How the birds are able to adjust for the real position of the prey is not understood. There is probably a considerable learning process involved, with many trials and many errors before the birds learn how to compensate between apparent and real depth, and different angles of view. Interestingly, Western Cattle Egrets *Bubulcus ibis*, which are herons that feed predominantly on terrestrial vertebrates and invertebrates, seem unable to successfully compensate for refraction. They have many more unsuccessful strikes when presented with submerged prey compared with the aquatic-feeding herons. It may be that Cattle Egrets could also learn to do this given sufficient trials, but it seems not to be part of their everyday repertoire.

There is evidence that Pied Kingfishers *Ceryle rudis* are also able to compensate for refraction when making plunge-dives after hovering above water to locate prey. Again, it is not known whether this is the result of much trial-and-error learning before the birds are able to develop general rules. Clearly, the closer the prey is to the surface the smaller the scope for error in estimating apparent position and depth. Also, a vertical dive will not require compensation for direction, although depth is distorted. Successful plunge-diving from directly above could bring success because the bird is bound to intercept the prey item, but it may not be able to predict accurately when it will hit it.

Skimmers

The three species of skimmers (*Rynchops* spp.) are found in the Americas, Africa, and India. They all employ the same special foraging technique to take fish prey from just below the water surface. They are renowned for flying above water on straight paths with their lower bill ploughing through the water. They are unique among birds in that the lower bill is blade-like and considerably longer than the upper bill (Figure 11.10). It is this lower bill that plays the key role in their foraging. When foraging, skimmers prefer calm water bodies: shallow lagoons, estuaries,

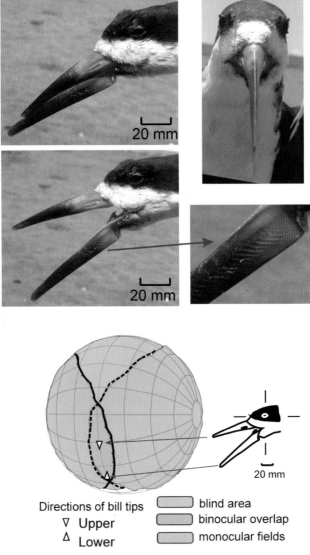

Directions of bill tips
∇ Upper
Δ Lower

⬭ blind area
⬭ binocular overlap
⬭ monocular fields

FIGURE 11.10 A Black Skimmer *Rynchops niger* showing how the blade-like lower bill extends well beyond the tip of the upper bill. The bones beneath are of equal length, it is the keratin sheath (the rhamphotheca) that extends so far forwards. It is thin and brittle and can be snapped off if a foraging bird hits a solid obstacle when ploughing the bill through the surface layer of water. The photograph on the right shows how thin the bill is, and the photo below shows the parallel ridges which extend along the mandible sides and are a mechanism that reduces drag when the bill is ploughed through water. The binocular visual field of skimmers (bottom) is narrow and vertically short; there is an extensive blind area above and behind the head. The binocular field is centred around the direction of the upper mandible but embraces the positions of both bill tips when the bill is open during foraging. It is not thought that skimmers use vision when foraging other than to stabilise their flying height above the water surface. The interception of prey at the surface is indicated by touch cues, and these trigger rapid closure of the bill to seize a prey item.

broad rivers, inshore waters. The waters in which they forage are often turbid, and in many situations the birds' foraging follows the tidal cycle, resulting in them regularly foraging in twilight or at night.

Perhaps not surprisingly, the foraging of skimmers does not appear to employ vision. While vision may be used to control the foraging flights, it is not used for the detection of prey items. Flights are usually in open habitats with few or no obstacles. It is probably not correct to say that skimmers 'locate' their food items, rather they are 'encountered', and their presence 'registered'. These prey encounters are recorded by the mandible as it slices through the water. This slicing is very efficient, adding little extra effort to the flight because not only is the mandible very thin, its sides contain small parallel ridges which reduce drag. Each encounter of the mandible with a prey item is signalled by some kind of tactile cue. This signal triggers a rapid snapping shut of the bill, the head momentarily ducks backwards beneath the bird and is then tossed forward holding the prey, and the bird carries on flying, often ingesting the prey as it does so.

Surprisingly, no evidence of tactile receptors of the kinds found in the bill-tip organs of other birds (Chapter 7) have been found in the mandibles of skimmers. The edge of the bill that cuts the water is as thin as a blade and there are no bundles of touch receptors along its edge. The bill is in fact disposable, and it often breaks off. These breakages occur because the birds frequently hit obstacles, such as floating logs, rocks, sand, and shingle. On such impacts the keratin sheath sometimes snaps off. The keratin sheath continually grows and is considerably longer than the bone that lies beneath. Because of wear and breakage, it is difficult to say what the typical length of a skimmer's bill is. The collection of skimmer skins and skeletons at the Natural History Museum shows the bills to be of many different lengths and shapes; some are blunt, some pointed, some rounded, all showing signs of wear and tear.

Because of this foraging technique it is unlikely that skimmers know what they are catching. Skimmers are another example of a species that forages blindly and has to verify its prey once caught. They frequently catch floating vegetation, small twigs, and other debris, but they rapidly discard these, presumably using touch and taste cues from inside the mouth to decide whether an item should be ingested.

Given that there are no touch receptors in their lower mandibles, the problem remains as to how skimmers actually register that their bill has encountered something just below the water surface. Two possibilities have been suggested. First, when something is hit the head is rotated forward by the impact, and it is this head movement that triggers bill closure. Second, hitting an object might stretch the muscles of the jaw, and receptors associated with the joint or the muscles trigger bill closure.

However prey encounters are signalled, the actual foraging behaviour of skimmers is very impressive. They are able to grab items while continuing to fly forward on a level and straight path. Foraging at low light levels or at night, when fish are more likely to be near the surface, is possible because the only visual requirement is for the birds to detect the water surface. There is even a suggestion that skimmers may attract fish. This is achieved by flying a short straight path with the bill cutting the water. This causes surface ripples which attract the fish, and the skimmer then turns through 180° and flies back along the same path to scoop up a fish.

Plovers

The final example of foraging at the water surface is provided by some species of plovers (Charadriidae). Plovers have short bills and are probably visually guided in their foraging, typically taking small invertebrates from muddy surfaces, although it has been suggested that some species could use hearing to detect the movements of invertebrates at or just below the surface (Chapter 6).

There is evidence that some plovers extend the immediate visual range of their foraging and may in fact use a variant of a flush-foraging technique. This comes from some intriguing findings about the vision and foraging of Blacksmith Lapwings *Vanellus armatus*. These are birds of sub-Saharan Africa where they are particularly noted for foraging for fish and invertebrates in the shallow water of pools, including their sometimes-muddy edges formed by the activities of mammals coming to drink. Foraging in these locations seems to be under visual guidance, with birds behaving stealthily or making short runs towards targets. Prey are taken with accurate bill pecks, and they are probably guided by prey movements at or just below the surface.

The visual fields of Blacksmith Lapwings show the typical arrangement of a visually guided forager. The bill is placed approximately centrally in the binocular field. This is correlated with their readily observed visually guided foraging behaviour. However, Blacksmith Lapwings also use a technique referred to as 'foot-trembling'. This is reported to occur in a number of other plover species. Foot-trembling involves the bird standing on one leg while the other is held out to the side and shaken or trembled to disturb the surface. When doing this a bird cannot see its foot, because it falls outside the visual field. It seems that this is a version of flush-foraging in which the trembling foot is used to disturb prey and force it into making an escape response, with the possibility that the prey moves into a location where it can then be detected visually and caught in the bill. Foot-trembling thus functions to extend the effective foraging area of a bird beyond the limits of it visual field.

Conclusion: foraging underwater and at the water surface

Foraging underwater or at the water surface often involves foraging at nocturnal light levels. The absorption and scattering of light by even clear water means that a bird that dives beyond about 50 m during daytime enters twilight. Because of suspended and dissolved substances, twilight levels may be reached at much shallower depths. Furthermore, many aquatic foragers are also nocturnal foragers, perhaps taking advantage of prey moving to shallower depths at night. Some aquatic foragers may be foraging in complete darkness, at depths where light levels are well below those of night at the surface. However, unlike in terrestrial nocturnal activity there seems to be no scope underwater for the senses of hearing and olfaction to complement vision.

Spatial information from vision underwater, even at high light levels, is perhaps best considered to be partial, or at least lacking in detail. From the perspective of a diurnally active terrestrial bird, underwater foragers would seem to suffer from a paucity of information. It is perhaps surprising, therefore, to find that underwater foraging by birds can be highly efficient. It is perhaps even more surprising that birds may rely on random encounters with prey or be guided towards prey by the minimal cues of small points of light from photophores.

As in the case of nocturnal foragers, the key to aquatic foraging often lies in a specialised prey capture technique. Such techniques allow particular prey to be taken guided by only minimal information. An amphibious bird will experience radically different worlds, and experience radically different perceptual challenges, as they oscillate between their life in air and their life below water. Their world view must change rapidly from one that is rich in information to one in which there is a paucity of information.

Birds which forage at or through the water surface rely primarily upon vision, but the water surface and the transparency of water pose their own set of sensory challenges. Clearly herons and kingfishers do meet these challenges very successfully. Their technique, however, may involve a large number of cognitive components, perhaps born of trial-and-error learning about the disparity between the real and apparent positions of objects when viewed through a water surface.

Among the skimmers it seems that tactile cues are their sole means of knowing when a prey item has been encountered, and that they rely upon random encounters with prey. It seems that random encounters with prey may also account for some of the foraging success of deep diving birds. Tactile cues are likely to signal these encounters, but the actual cues are unknown. In skimmers they serve to trigger very rapid bill closure, but they do not tell the bird what it has encountered, so mistakes are made, inedible items are caught and have to be discarded.

While the foraging of birds underwater or at the water surface usually relies on sparse information, it is nevertheless highly successful. It supports the lives of a wide variety of species whose populations number in the millions. However, this reliance on sparse information makes these birds particularly vulnerable to problems caused by human artefacts, especially fishing nets, placed in their underwater environments. These vulnerabilities will be discussed from a sensory ecology perspective in the final chapter.

Chapter 12

A sideways look through birds' eyes

It is time to stake stock, to stand back a little from the detail and subtlety that has been explored in previous chapters. Time now to consider some general lessons that investigations into the sensory ecology of birds have revealed, and to consider some applications of sensory ecology to conservation issues, especially the problems of collisions with obstacles and entrapment in nets. Why a sideways look? Simply to emphasise that for birds their visual worlds are much more extensive than ours and that their regions of most acute spatial and colour vision project sideways from both sides of the head, not forwards. Humans are the unusual forward-looking animals. For birds, looking sideways is the norm.

A bird is a bill guided by an eye

The phrase 'a bird is a wing guided by an eye' was coined almost 80 years ago. It was an early attempt to characterise the basic biology of birds from a sensory ecology perspective. The phrase drew its inspiration from the idea that the all-embracing feature of bird biology is flight. It drew on the idea that the key features of birds can be explained by considering them, first and foremost, as 'flying machines'. This provided a powerful explanatory framework for thinking about birds as animals with low body mass but whose muscular, respiratory, and digestive systems were capable of generating high power output. If the ability to fly could be considered to have been so dominant in shaping these core aspects of bird biology, it was a small step to make the assumption that the control of flight would also make particular demands on the information that birds gather from their environment.

It now seems, however, that this assumption about the importance of vision in flight may have been misplaced. As more has been learned about the sensory ecology of birds, it seems that controlling flight is in fact not the prime driver of vision. The key driver of bird senses is now perhaps best regarded as stemming primarily from the demands of foraging, that is the detection, acquisition, screening, and ingestion of food items. The second key task that has shaped bird vision is the detection of predators. Furthermore, as explained in Chapter 5, the tasks of foraging and predator detection probably make competing demands for information. From this perspective, the informational demands for controlling flight are achieved within the requirements set by the sensory challenges of foraging and predator detection.

In light of this, it now seems more appropriate to regard a bird as 'a bill guided by an eye'. This is because the bill is the prime tool with which birds interact intimately with their environment and is used primarily to gain food items, requiring the bill to be brought to the right place at the right time to seize an item. This is an exacting task requiring precise and accurate information. For some species, particularly some of the birds of prey, it might be better to suggest that they are primarily 'feet guided by an eye', since it is the feet that do the crucial act of prey acquisition.

This change in the shorthand way of describing birds may seem rather superficial, but it does go to the very core of 'what is a bird?' This thumbnail description of a bird is important because it changes how we might think about birds in an everyday sense. It frames the narrative. It indicates that the main task of a bird, day in, day out, is getting its bill (or feet) to the right place at the right time, over and over again. It not only sums up what birds actually do; it also makes it clear that they have to get it right. These tasks, and their informational control, are subject to constant selective pressure. It is no wonder that in foraging, getting the right information is as important as having the right shape of bill. Making a sensory-based error when flying may be important, but it is not as important as being unable to explore the environment at close range and use your bill to acquire food.

Seeing a bird as 'a bill guided by an eye' also has important utility. It provides a framework that gives a good grasp of basic bird biology, and it also provides an appropriate framework for thinking about how birds in general interact with their environments. Characterising a bird in this way also has application in some key conservation issues. For example, as explained later in this chapter, it can help in understanding the circumstances that lead some birds to be vulnerable to collisions with large obstacles or what makes them vulnerable to entrapment in fishing nets, problems which result in the death of many millions of birds annually.

Differences in sensory capacity between species

In the early chapters it was shown that sense organs have the same relatively simple structures across all species of birds. However, the key components of sensory organs, and especially the arrangements of the sensory receptors that they contain, are highly flexible. Because of this, marked differences can occur between species in all sensory capacities. For example, different eyes can produce images with different characteristics of brightness, quality, and extent. Furthermore, those images can be analysed in a seemingly endless number of different ways depending on the densities and distributions of different photoreceptor types across the retina.

This flexibility extends to individuals and means that natural selection has a highly diverse base upon which to act in the evolution of sensory capacities in different species. Some of those differences in sense organs can be large and obvious, some can be very subtle. They all attest to the fact that senses can become tuned to both large and small differences in the sensory challenges faced by different species in different environments.

This makes generalisations about sensory capacities rather problematic. As an example, consider the diversity of bills found across a range of bird species. We do not expect these bills to have the same properties and functions, and we would not simply assume that a 'bill is a bill' and functions more or less in the same way in all species. There is much fascination in the details of bills: their shape, length, and strength. There is also much fascination in understanding how these properties are tuned to different mechanical functions that match different foraging tasks.

The same is true of sensory capacities. Certainly, general principles can be discerned, and we may be able to assume that certain species are able to extract similar information through a particular sense. However, we should be wary of applying generalisations without considerable thought, and preferably investigation. This applies to all of the senses, not just vision. While there is a commonality in the types of receptors for taste, olfaction, and touch across bird species, their actual numbers and distributions vary greatly between species, and this results in a fascinating diversity in the information that different species have available to them.

The key conclusion is that all senses vary between species and that differences result in species being able to extract different information from the same environment. This might be information about objects at a distance from the bird, or information about objects and substances in touch with the body or even within it, especially within the mouth. Not only are there different birds'-eye views, there are also different batteries of information available to each species about tastes and smells, different degrees of information about sounds and touch, and perhaps different degrees of information about the earth's magnetic fields.

Complementarity and trade-offs between the senses

When trying to explain the ability of different species to exploit the resources of different environments, it was argued that information from a single sense could only rarely account for a particular behaviour. It seems that there are always complementary relationships between different senses in the conduct of tasks. For example, for an owl to capture prey beneath a woodland canopy at night requires information from different senses to be used in complementary ways. No one sense dominates. There is not a single gee-whiz super-sense explanation. Although one sense may dominate a particular behaviour in a sequence, the whole task cannot be completed without information gained through other senses.

An owl may rely upon hearing to locate and even pounce onto prey, but vision will have been used to locate the hunting perch and to locate a place to take the prey once caught. A kiwi may use smell, touch, and hearing in a complex way to provide information when searching the forest floor for earthworms, but the kiwi will probably also use vision, albeit of low resolution, for its general orientation within its foraging area. Clearly there are limits to all sensory capacities – the lowest light level that can be detected, the quietest sound that can be located – but these limits are often met or circumvented by employing complementary information from another sense.

A paucity of information and the role of cognition

Another striking conclusion from the later chapters has been the evidence that birds often rely upon very sparse information to guide key behaviours. In fiction and religion, the 'all-seeing eye' (the eye of providence) is a powerful idea. It crops up everywhere, even on a US one-dollar bill, but it is just a fiction. Eyes, like all senses, have limits and are compromised by the conditions of the environment in which they operate.

Many of the key behaviours of birds, and indeed many of our own key behaviours, are conducted when eyes can be far from all-seeing, when spatial information is gross rather than detailed, providing only a relative paucity of information. To an observer recording the behaviour, however, it might be concluded that a bird does have all-seeing eyes, at least eyes always capable of providing a high degree of spatial detail. Behaviours that are guided by a paucity of information are possible because the animal is able to interpret such sparse information in a functionally useful way. That is, there is often a significant cognitive component that underpins the behaviour when information is sparse. The bird either knows about the specific place and the structures that occur there, or has some kind of knowledge about general properties of the environment that enables it to exploit limited information in a meaningful way.

This can be seen in the examples of how sensory information is used in the guidance of nocturnal and underwater behaviours. Perhaps the most extreme examples are found in birds that forage in highly turbid waters or at depth in waters with very low ambient light levels. In these situations, we can be confident that the birds do not have available to them highly detailed spatial information. However, they are able to act as though they did because specific or general knowledge about the underwater environments allows them to interpret what information there is in a functionally useful way. This situation may well also occur at higher light levels, but in such circumstances it is less easy to know what information an animal has available.

One example, which brings home to us that animals do behave regularly, sometimes habitually, guided only by sparse cues, comes from ourselves when driving a motor vehicle, especially at night. When I am behind the wheel, I find it very easy to frighten myself by reflecting on the information that I have available from moment to moment as I travel at speed. I urge every reader to do the same – it might make you a safer driver.

Can you be sure that you are travelling within the information that you actually have available at any one instant? Considering this question is especially frightening when travelling at 100 km/h at night. Even with headlights, what information do you actually have that allows you to travel at that speed? Are you really aware with any certainty of what lies ahead? Are you, in fact, relying upon partial information supported by your knowledge of the nature of roads? Are you actually picking up only sparse cues and making the prediction that the world will continue for the next 100 m as it has done in the past? Is your driving relying upon the very limited information provided by conventional, highly stereotyped road markings and signs, and lights on other vehicles?

The most unsettling question to ask is, if a deer wandered into your path would you even see it, let alone be able to avoid colliding with it? Probably not, for the current estimated number of vehicle–deer collisions each year on UK roads is 42,000; such collisions are not rare events. Most of these collisions fall within a class of road accidents that are described as 'looking but failing to see'. That is, drivers are looking and attentive but the task of seeing the wandering deer is often beyond their perceptual limit.

This means that people are behaving beyond the information they immediately have available, yet they are able, on the majority of occasions, to complete the task of driving because the sparse information that is available can be interpreted successfully in commonly occurring circumstance. If roads change, for example if there is a hazard or road works ahead, it is necessary to seemingly overload drivers with information to warn them: an abundance of traffic cones, additional lighting, advanced warning signs. These are necessary to make drivers take stock and realise that the world is going to change, that the interpretation of minimal cues at 100 km/h will not be sufficient to negotiate the hazard.

The generality of this kind of interpretation of minimal cues to other situations and other species is, of course, open to debate. But it does provide a useful framework for thinking about the limits of sensory information, especially visual information, which is so markedly influenced by ambient light levels. It also provides a framework for considering how cognition, general and specific knowledge of environments, allows actions that would otherwise not be possible if they had to rely on more detailed information. As argued in previous chapters, this certainly seems to occur in birds in a number of situations. It probably also applies to many, if not all, behaviours that take place in environments in which sensory information can be markedly restricted.

Trade-offs or compromises within a sense

The early chapters described many examples of how senses have limits, how achieving one particular capability has consequences for others, especially in multifaceted senses such as vision. As mentioned above, an all-seeing eye is just fiction. A particularly informative example of a trade-off within a sense is found in the high-resolution vision of large birds of prey. It is informative because it shows how gaining high performance in one attribute can affect quite another aspect of vision, with important consequences for understanding some specific behaviours.

Detailed investigations have shown that the highest acuity yet recorded in any vertebrate eyes is found in the larger birds of prey, particularly eagles and Old World (*Gyps*) vultures. We are usually impressed by the fact that an eagle is likely to have visual acuity five times that of the average middle-aged person, or at least twice that of the average keen-eyed young birdwatcher. It is also impressive that an eagle's resolution is many more times higher than those of the birds which live alongside it.

These differences in resolution are clearly understood in terms of eye optics and retinal structures, and they are also readily understood as having important behavioural consequences. The prime behavioural advantage is that objects can be detected at great distances. Certainly, an eagle is able to detect a target at a much greater distance than other birds or a human observer. The functional driver for such high acuity is the need to detect prey or carrion at long distances, allowing an eagle to effectively forage for prey that is sparsely distributed over large areas of open habitat.

It is exciting to see a vulture or an eagle starting to descend from great height apparently towards an item which we cannot hope to see at the same distance. However, achieving such high acuity has costs. The eye has to be large and therefore heavy and metabolically expensive to support. Achieving high acuity also means that the sun should never be imaged on the retina. Smaller-eyed birds with lower acuity can tolerate an image of the sun being formed on their retinas. Without this, they could not achieve comprehensive coverage of the world about them, essential for detecting predators. However, an image of the sun falling on the retina in a large-eyed bird cannot be tolerated, since it would immediately negate the high resolution of the eye. This is because the image will itself become a bright light source, scattering light around inside the eye and resulting in glare and a subsequent loss of high resolution. It is for this reason that large-eyed birds have 'sunshades', in the form of brow ridges and long 'eyelashes'. The cost of these brow ridges is that they severely restrict the bird's field of view, resulting in a large blind area above the head that extends into the upper part of the forward field of view. In itself this is not a problem, since as a top predator an eagle will have few, if any, predators that it must watch out for when flying.

The extensive blind area becomes a problem when an eagle or vulture is foraging and searching the landscape below. The head must be pitched forward, with the result that the flying bird is not able to keep a lookout ahead. It is

FIGURE 12.1 A *Gyps* vulture foraging on the wing. The head is pitched forwards as the bird searches the terrain below. There is a large blind area above the head that results from 'sunshades' (see Figure 5.13). The blind region extends sufficiently forwards that with its head pitched at an angle of 60° this bird will be blind in the direction of its travel. This renders the bird vulnerable to collisions with large man-made objects, including power lines and wind turbines, that intrude into its naturally open airspace.

effectively blind in the direction of travel, rendering it vulnerable to collisions with objects that intrude into the open airspace. Of course, such obstacles are rare in the natural environment, but human artefacts such as power lines, transmitter towers, and wind turbines do intrude into the open airspace, and these birds are particularly vulnerable to collisions with them (Figure 12.1).

This trade-off, between large eyes providing high acuity, and restrictions of the visual field to avoid imaging the sun, is found in many large-eyed bird species including hornbills, bustards, albatrosses, and even ostriches. All of these species have 'sunshades', especially prominent eyebrow ridges, suggesting that keeping the sun out of their eyes is a problem common to them all. Crucially, in certain species it results in them becoming vulnerable to collisions when humans encroach upon their environments.

The problems of collisions and entrapment

Power lines, fences, communication masts, and buildings have long been recognised as posing major problems for certain bird species (Figure 12.2). In some instances, birds collide with objects that may be partially obscured by vegetation, for example fences in woodlands, but collisions typically occur with structures that extend prominently into the open airspace, above surrounding vegetation. These 'hazards' appear very conspicuous to humans, but apparently not to some birds. Some of the collisions occur at low light levels, but many occur in full daylight. Birds from a wide range of species are also recorded as colliding into panes of glass, and many species of diving birds become entrapped in fishing nets. Many birds also die as a result of being hit by fast-moving objects such a road vehicles, aircraft, and trains.

These instances of bird mortality through collisions and entrapment are not trivial. It has been claimed that collisions with static and moving human artefacts are the largest unintended human cause of avian fatalities worldwide, and up to 100 million birds are estimated to be killed annually in this way. There is evidence that collisions with masts and power lines may threaten the survival of specific populations, or even the survival of certain endangered species. Species known to be affected in this way include White Storks *Ciconia ciconia*, Blue Cranes *Grus paradisea*, Ludwig's Bustards *Neotis ludwigii*, and Kori Bustards *Ardeotis kori*. Collisions involving large raptors, eagles and vultures, are the focus of intensive

FIGURE 12.2 Human-made collision and entrapment hazards for birds. Large and apparently very obvious static objects (top row), as well as fast-moving vehicles of different sizes (middle row), and smaller static structures such as fences and glass panes, are all collision hazards, killing many millions of birds annually around the globe. Fishing nets pose very obvious entrapment problems for all species of birds that dive to forage. The problems with all types of collision hazards have their roots in the limitations of the sensory information that birds have available to them. They pose sensory challenges which the birds' behaviour and sensory systems have not evolved to cope with.

research work, because of their vulnerability to collisions with wind turbines. That diving birds get caught in fishing nets is perhaps not so surprising, since nets are designed to be inconspicuous, but collisions with large apparently obvious static objects seem surprising.

Collisions with moving objects, such as motor vehicles, trains, and aircraft, would seem to pose a different set of perceptual challenges for birds. This is in part due to the high speeds with which these objects can move. Structures in the open airspace, fishing nets in open waters, and large fast-moving vehicles are very recent challenges from an evolutionary perspective. It is worth asking whether these problems have similar causes, and whether they have similar mitigations? Can humans design a way around these collision hazards?

Unfortunately, it seems that there is not a single cause or a single answer to these problems, but they can be understood by taking a sensory ecology perspective. For example, many collisions with static objects may occur because birds are simply not detecting the target. A fast-approaching vehicle, on the other hand, may be interpreted by a bird as an approaching predator. This initiates a standardised predator escape behaviour which in many circumstances allows insufficient time to escape successfully. In essence, it is the speed of the approaching vehicle that kills because the apparent predator is travelling at speeds in excess of those of the bird's natural predators.

Collisions with static objects

Measures to reduce the probability of collisions with static objects have usually involved marking the obstructions with devices which are designed to increase the chance that they will be detected from a greater distance than the actual hazard. The assumption is that the hazard itself is below the limit of visual resolution within the flight avoidance distance of many bird species. For example, power lines have been marked with objects such as reflective balls, flapping flags, and wire coils. However, despite more than 30 years of using static markers on power wires, the probability of mortality caused by power-line collisions remains high for certain species.

Until recently, solutions for reducing collisions have been based on a human perspective of the problem. Put simply, it has been a matter of trying to find solutions to bird collision problems by making the perceived hazard more conspicuous to human observers. Furthermore, work on the development of hazard markers has been constrained by the need to find solutions which have low initial cost, are easy to apply, and are easy to maintain. What is likely to be conspicuous to a bird flying in a particular environment or situation has rarely, if ever, been considered. Solutions have been off-the-shelf, based on local availability of devices or traditional practice, and on what engineers can see.

Sensory ecology factors that predispose towards collisions

Can there be a sensory ecology-based solution to collisions, or at least a sensory ecology-based set of recommendations for collision mitigations? To answer those questions requires a brief summary of the reasons why bird species may find particular structures or situations hazardous. Such a summary may present few surprises to someone who has now read this book, but it is worth summarising because it emphasises that solutions to collisions based on a human perspective are likely to be misleading.

First, it must be acknowledged that birds live in quite different visual worlds to that occupied by humans. After reading this book I should hope no one would find difficulty with this statement, but surprisingly it can be a revelation to an engineer charged with the task of reducing collisions with power lines or wind turbines.

Second, in flight some birds may be blind ahead of them. Turning the head in both pitch and yaw to look downwards, either with the binocular field or with the central part of an eye's visual field, may not be unusual. But in a number of collision-prone species this can result in them being blind in the direction of travel.

Third, frontal vision in most birds is not high-resolution vision; highest resolution occurs in the lateral fields of view, the sideways view. So, when calculating whether a hazard or a marker device will be visible head-on to a bird, an estimate based on the highest acuity (lateral vision) is misleading. It is necessary to estimate what the bird's acuity might be in the direction of travel, not the bird's highest acuity. Most birds probably employ lateral vision for the detection of conspecifics, foraging opportunities, and predators. Attention to these may be more important for a bird than simply looking ahead during flight in the open airspace.

Fourth, birds in flight may be predicting that the environment ahead is not cluttered. Even if they are 'looking ahead' they may fail to see an obstacle since they may not predict obstructions. Here the analogy with human car driving may help to understand the problem.

Finally, it should be acknowledged that birds have only a restricted range of flight speeds. Unlike the car driver, birds cannot adjust their speed. This means that they cannot adjust the rate at which they gain information as they move through the world. We can readily adjust the rate at which we gain information so as to ensure safe progress. Careful drivers will do this when the sensory challenges of the environment change due to reduced visibility caused by rain, mist, or lower light levels. Each bird species, however, has a narrow range of optimal powered-flight speeds. This speed is dictated by the bird's wing loading: birds with high wing loading fly faster than those with lower wing loadings. Whatever their optimal speed, birds cannot slow down to take a closer look at what lies ahead. This means that hazards have to be detected at a sufficient distance and time in order to take avoiding action without changing speed.

Sensory ecology and collision mitigation

Although birds cannot be guaranteed to be looking at, or attending to, an obstacle that extends into open airspace, it is still valuable to employ markers to increase their conspicuousness in key situations where collision rates are high. Markers that are used to draw attention to the obstacle should, however, be as conspicuous as possible, perhaps in one sense paralleling the apparent overload of conspicuous and repeated information that is used to warn car drivers of a change to the road ahead.

The warning stimulus should ideally be physically large, well in excess of the size calculated to be detectable at a given distance based on acuity measures, and it should incorporate movement (Figure 12.3). These features should increase the chances that a stimulus will be detected at a distance sufficient for a change in flight path to be initiated. This recommendation takes account of the idea that

FIGURE 12.3 A new design of bird diverter deployed on high-voltage power lines. It rotates freely in the wind about a pivot joint and combines large size, high contrast, and movement. These features are designed to make the diverter conspicuous to a wide range of flying birds based on known visual acuities and flight speeds. Under both bright daylight and twilight these features should enable collision-vulnerable birds to see the diverter at a distance sufficient for them to divert away from the hazard.

forward vision in birds may be tuned primarily for extracting information from optic flow rather than static stimuli (see Chapter 5), while estimates of acuity typically refer to the highest performance of spatial resolution, which occurs in the lateral, not the frontal, fields of view.

It is worth recalling that foraging birds will complete a wide range of key behaviours guided only by minimal cues, or a paucity of spatial information. It is highly likely that birds are similarly prepared to fly in what they predict will be open airspace when there is a paucity of spatial information available to them. This might arise through reduced visibility caused by mist, rain, or low light levels.

A warning structure that is likely to be the most conspicuous under all possible viewing conditions (e.g. reduced light levels, reduced contrast due to mist or rain) should simply employ a high-contrast black and white pattern. This ensures that the warning structure reflects highly or absorbs strongly across the full spectrum of ambient light, regardless of the ambient light regime. The degree to which a coloured target is conspicuous depends on the spectral characteristics of ambient and background illumination. These can vary markedly with situation, time of day, and cloud cover. Furthermore, colour vision is lost at twilight and night-time light levels, so if the warning is to remain conspicuous over a wide range of natural light levels, it is best to use black and white.

Collisions with obstacles in the open airspace may be as much a perceptual problem as a visual one. Therefore, if possible, solutions should be found that alert birds well in advance: their attention may need to be primed just as much as the car driver's when approaching a hazard. However, what constitutes a warning or alerting stimulus may be difficult to determine and is likely to vary with species, but auditory as well as visual warning stimuli might be effective, since they may generally stimulate renewed or increased attention to the environment and the birds may search generally for a possible hazard.

Diverting or distracting birds

At locations where collision incidents are frequent it may be more efficient to divert or distract birds from their flight path rather than attempt to make the hazard more conspicuous. For some collision-prone species, a signal on the ground may be more important than a signal on the obstacle. These may take the form of foraging patches, conspecific models, or alerting sounds positioned a suitable distance from the hazard. The rationale is that birds might be induced to land and then be more attentive to possible obstacles as they take off again into the airspace, rather than simply flying through it with reduced attention to what lies ahead.

A belt and braces approach might be the best option in certain situations and with particularly collision-prone species. Conspicuous hazard markers, combined with manipulation of the sites in the vicinity of the hazard, would always seem to be the best option. For low-tension power lines which are at only a modest height above ground, planting barriers of trees on either side of the line should force birds to fly higher and avoid the hazard completely. Such environmental manipulations are likely to have general ecological benefit, but in many instances they may not be possible, as those responsible for the hazardous structure may have limited control of the land in the vicinity of the hazard.

Bespoke collision solutions

There is unlikely to be a single effective way to reduce collisions for multiple species at any one site. Warning or diversion and distraction solutions may need to be tailored for particular target species. Solutions may need to take account of the foraging ecology and social behaviour of the species as well as its visual capacities in order to understand why it flies in the open airspace at particular locations and renders itself vulnerable to collisions.

Collisions with moving vehicles

Can anything be recommended from a sensory ecology perspective to mitigate bird collisions with moving vehicles?

The problems surrounding bird–vehicle collisions seem to stem primarily from the behavioural responses of birds rather than a specific sensory problem. The sensory ecology factors described above that affect static collisions will apply to moving-vehicle collisions, but there is also an additional perceptual problem. It seems that birds interpret a fast-approaching vehicle as a predator. This typically triggers an escape response when the 'predator' appears to be at a set distance, regardless of its speed. When faced by a potential predator, birds seem to employ some kind of 'distance rule' for initiating an escape response, and this is also applied by birds to oncoming vehicles. Therefore, vehicle speed is crucial in determining whether a bird can escape.

Experiments with Brown-headed Cowbirds *Molothrus ater* showed that their escape response was usually initiated less than 0.8 seconds before collision (the time needed for escape) when vehicles approached at speeds greater than 120 km/h. This is a late response and results in a high risk of collision when birds are faced with highway traffic. Anecdotal observations would suggest that other bird species behave in the same way when confronted by fast-moving vehicles, and that they too apply a distance rule before making an escape. However, different species may initiate their escape responses at different distances. For example, it is well established that different species have different flight initiation distances when approached by actual predators.

Mitigation measures based on these findings are, however, rather few. As in the case of static collision hazards, the reduction of fatalities primarily requires location-based management of the environment to reduce the encounter rate between birds and the hazard. Increasing the conspicuousness of the moving hazard, beyond the current practice of employing lights on road vehicles at all times, may not be possible. It is clear that speed does kill and lowering the speed of vehicles would seem to have greater benefit than any efforts to increase the general conspicuousness of vehicles. Next time when you are driving and you encounter a pheasant wandering in the road, remember that it probably views your vehicle as a predator and it is unlikely to move until you draw closer than a certain distance. The obligation is on you to slow down to avoid the collision.

Gillnets and diving birds

The fact that birds are caught in fishing nets, especially gill nets, is not surprising. These nets hang in the water at various depths for many hours and can be kilometres in length. After all, the nets are designed and set to trap animals. Gillnet bycatch of seabirds is a worldwide problem which results in the deaths of at least 400,000 birds annually. Nets are deployed at a huge scale all around the globe. The annual rate of bycatch is thought to be unsustainable for some bird species, and there is evidence that in some localities it has resulted in severe reductions in the numbers of breeding birds. Gillnet bycatch is not unique to birds. It is a well-established issue for other animal groups including sea turtles (Chelonioidea), pinnipeds, cetaceans, and blue-water fish (tunas *Thunnus* spp. and billfish Istiophoridae and Xiphiidae).

It is recognised that there is an urgent need to reduce all of this bycatch. However, in order to obtain support and adoption by the fishing industry of any mitigation measures, it is desirable not to reduce the efficiency of nets for fishing. This may be an impossible task. Temporary closures of fisheries have been shown to be effective in managing the impact of gillnets on specific populations, but such measures may be difficult to establish and enforce, and they do not actually solve the problem of bycatch per se; rather they just limit when and how frequently it occurs. Can a sensory ecology perspective throw any light on possible technical solutions to the bird gillnet bycatch problem? Can nets be made less efficient at catching birds while maintaining their ability to catch target fish?

Bycatch bird species

As with aerial collisions, bird species that are prone to entrapment in gillnets come from a range of orders and families. They include species which differ markedly in their foraging ecology, suggesting that the problem of gillnet bycatch may derive from rather broad and general sensory ecology factors rather than specific factors which predispose only certain species. The important conclusion from recent reviews of bycatch is that practically any bird which forages where gillnets are deployed is liable to be caught in them. As far as birds are concerned gillnets are a catch-all device, and no particular diet or foraging strategy seems to make birds particularly prone to entrapment in gillnets.

Vision in gillnet bycatch species

Chapter 11 described the sensory ecology of foraging underwater and indicated that diving birds must frequently forage under conditions when they have available to them only gross spatial detail. They may in fact not be able to see high detail once their eyes are immersed. In light of this it is not surprising that gillnets are a hazard to diving birds. Unless near the surface in clear water and in bright daytime light conditions, gillnets will not be visible until the birds are very close to them, and under many circumstances the nets are undetectable.

Warning and distracting

The solution to gillnet bycatch would thus seem to require either that the presence of nets is signalled to the birds in some way, and that the birds retreat from them, or that the birds are decoyed away from the vicinity of nets. Decoying is difficult because, unlike power lines and wind turbines, the locations of nets change almost on a daily basis. They are not fixed structures, and their surrounding habitat features cannot be manipulated.

Making the actual nets more visible is unfortunately not an option. Nets need to be invisible to target fish species but at the same time more detectable to bycatch species. These two requirements cannot be reconciled. But there is a further problem in that if the nets are detectable to birds, they could in fact attract the birds to them, for they may be curious about them.

Another possible solution is to make the general position of the nets conspicuous, in the hope that the birds will keep away. Two approaches have been tested: lights along the top line of the nets, and conspicuous panels placed at intervals along the net surface.

Net lights

It may seem attractive to use light sources placed upon the nets to draw attention to them and/or to make them more visible. Indeed, such devices have been employed with some success aimed at reducing gillnet bycatch of turtles at night. However, this might not be so advisable with birds because of the effect of bright lights on the adaptation of the retina to the ambient light. When a bird is foraging either at night or at depth, its retina will have a high degree of dark adaptation. Exposure to a light source within the twilight–daylight luminance range will produce a rapid reversal in adaptation. This will result in impairment in the retina's ability to resolve detail, at least within a portion of the visual field, and this impairment will last for a period considerably longer than the brief exposure

FIGURE 12.4 Examples of the types of warnings that have been trialled to deter diving birds from approaching gill nets. Trials were conducted in a fishery in the Baltic Sea. The high-contrast panels regularly spaced along the net surface were of large size. They should have been detectable by ducks at sufficient distance for the birds to take avoiding action under a range of natural light levels and water turbidity. In other trials lights were spaced at regular intervals along the headline of the net; in some trials the lights were a constant green, and in others flashing white lights were used. Compared with control trials, none of these measures was found to reduce bird bycatch in gill nets. (Modified from an original drawing by Rob Enever.)

High contrast net panels

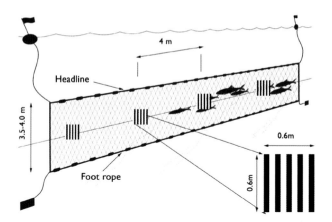

Constant green net lights

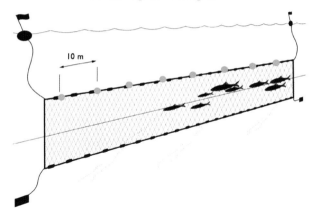

Flashing white net lights

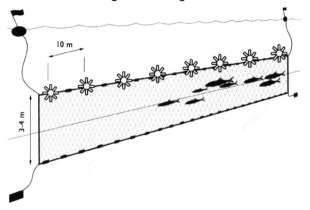

to the light. Therefore, exposure to a light is likely to increase the probability that parts of the net not immediately illuminated will be less visible.

Net lights have been tested in two different parts of the world, involving different bird species. One study, conducted on the Pacific coast of South America, showed a reduction in the bycatch of cormorants and turtles when nets were equipped with lights strung out along the top line of the nets. The other study, conducted in the Baltic Sea, showed no effect of the lights on the bycatch of sea ducks. Thus the jury is out, and more studies are required (Figure 12.4).

Warning panels along the net surface

To avoid the problem of disrupting the dark-adapted state of the bird's retina, warning panels incorporating high-contrast black and white patterns have been tested. These were designed to be conspicuous over a range of light levels and water turbidity. They should have been conspicuous to the target species, mainly sea ducks, over a distance of many metres, so alerting the birds to the hazard of the net. The results of trials were interesting but disappointing. On the whole the panels were not effective, making no difference, but in one study they seem to have resulted in slightly higher bycatch involving one particularly catch-prone species, Long-tailed Ducks *Clangula hyemalis*. This negative result was important in that it suggested the panels were detected by the birds underwater, but the birds' behavioural response was not one of avoidance. Rather it was one of curiosity or investigation, which increased their chances of being caught in the net. What this indicates is that the sensory ecology of the situation is understood but the behavioural response of the birds is not. Clearly, we need to know more about exactly what the birds are doing when they are foraging underwater.

Collisions with glass

Collisions of birds with glass panes has long been recognised as a problem of significant scale and was first quantified as a hazard to migrating birds, particularly passerines, 40 years ago. It has more recently been estimated that bird mortality due to collisions with glass could involve millions, maybe billions, of birds annually worldwide. If this is true it is a bird mortality problem whose total magnitude eclipses the combined effect of all other collisions and entrapment, although each hazard type has its impact primarily on different suites of species.

The problem of collisions with glass arises from two different properties of a glass pane: transparency and reflection. Because it is transparent, birds may simply see what is beyond the glass and attempt to fly to it. Because it is reflective, birds may perceive the reflection as a space into which they can fly, so in effect they attempt to fly into what is behind them. In either case the birds are unaware of the hard glass surface.

Mitigation measures

In response to this collision problem mitigation measures are frequently employed. They usually involve devices to disrupt either the transparency or the reflective properties of the glass. This is achieved by placing structures before or on the glass surface. Self-adhesive patterns to apply to glass are widely available, but it seems unlikely that they have reduced the incidence of collisions. More recently, architecture-based solutions have been proposed in which the glass surfaces are either recessed deeply or disrupted by blind-type external structures.

There has been little experimental investigation of the efficacy of such measures, owing to the difficulty of replicating under controlled conditions the types of situations that occur in the field. As in the case of gillnet bycatch reduction, the aim of any mitigation is to reduce collisions while still maintaining the essential properties of the hazard that are of benefit to humans. People are reluctant to forgo their large windows, much in the same way that fishermen are reluctant to forgo any of their fish catch.

Patterns on glass surfaces

The experimental work that has been conducted suggests that patterns applied to glass have limited efficacy unless they occur at such a density that they disrupt the essential benefits of glass to humans. Basically, they must disrupt the mirror effect of a glass surface. Patterns must be repeated right across the width and height of a glass panel and applied to the outside surface. Randomly placed designs were found to be ineffective unless applied at high density. These include the popular silhouettes of birds of prey, widely marketed as a means of reducing window collisions. Vertical stripes ideally need to be no more than 10 cm apart, and are more effective than the same pattern in horizontal orientation. Adding colour to the patterns makes them no more effective than black or white patterns.

One intriguing solution that had been proposed to this impasse was the use of UV reflective or absorbing markers on glass. UV marker pens and self-adhesive transparent, but UV reflective, strips have been readily available and have been marketed as a way of reducing bird strikes. The idea behind this is that since humans are not sensitive to UV the application of UV marks would not be detected by humans while they would be detected by birds, at least by passerine species which have relatively high sensitivity to UV light.

Careful modelling of the sensitivity of birds' eyes to UV light, the reflectivity and absorption of UV strips, and the amount of UV light available in the natural environment shows that such markings might indeed be visible to passerines but only under high UV lighting. They would not be visible to non-passerines even under high UV. Furthermore, the amount of UV reflected from the images of foliage and other structures in the area immediately surrounding panes of glass will frequently render the UV patterns undetectable. Thus, a simple UV-based marker solution does not seem to be viable other than under high natural light conditions. Flooding UV reflective or absorbing patterns with UV light to make them more conspicuous to birds would not be viable because of the potential hazard to human vision.

The sensory ecology of collisions and entrapment: conclusions

It is clear that efficient and convenient solutions to these problems have yet to be found. A sensory ecology approach has allowed a more careful analysis of the problems than was hitherto applied. Unfortunately, much still needs to be known. These problems and their solutions are where sensory ecology meets ethology. It may now be possible to understand the problems from the sensory perspective of birds. However, it is still not possible to be sure how birds will react when the hazard, or its warning signs, are detected.

A bird apparently treats the looming car as a predator. This is a response which is quite understandable. The pheasant that lingers in the road as a car approaches is not acting in a stupid way; it is acting in a way that has been appropriate for its species for thousands of generations. The eagle, vulture, or crane failing to keep a lookout ahead is also employing strategies that have evolved to provide efficient foraging for many thousands of generations. It is the nature of the hazard which humans have introduced in a handful of years which is the problem, not short-comings in the behaviour of birds or in their sensory systems.

Although incomplete and partial, knowledge of the sensory ecology of birds provides a way of thinking about behaviour from the perspective of the birds themselves rather than the perspective of a human looking on. The idea that birds might live in different sensory worlds to our own was first discussed by Greek philosophers more than 2000 years ago. Modern science has given us overwhelming evidence that this is true. It is now time to use that evidence to think about birds from a fresh perspective. In particular, we need to understand the world from the birds' points of view. We need to look through the birds' own senses at their ecology and behaviour, and thereby understand and solve the problems that human activities have created for them.

Appendix

Visual acuity in birds

Visual acuity is the most convenient way of comparing the spatial resolution of eyes. Acuity gives an estimate of the smallest detail that an eye can detect, and the table presented here allows ready comparison between the 52 species of birds (from 13 orders and 25 families) for which acuity is available. These are estimates of the highest resolution of an eye when light levels are within the upper levels of the daylight range.

Acuity can be determined using three main methods, and this table combines estimates of acuity using all three. The primary method is acuity measured using behavioural techniques, as described in Chapter 2. This is the preferred method because it is based on what it is known that the birds can actually see under controlled viewing conditions. Acuities measured in this way are shown as 'behaviour' in the table. Two other methods use knowledge of the bird's retina combined with estimates of the focal length of the eye. These methods use data on the spacing of either the retinal ganglion cells, or the photoreceptors cells, in the region of the retina where cells are most closely packed. This is typically at a fovea or another recognised area such as a linear band (see Chapter 4). This method of estimating maximum resolution cannot take account of light levels and their effect on resolution, and it does require an accurate measure of the focal length of the eye. Acuities based on these measures are shown as 'ganglion' and 'photoreceptor' in the method column of the table.

In the table, resolution has been rounded to the nearest decimal point and is a mean value published for a species. There may be considerable differences between individual birds within a species sample, no matter what method is used to determine acuity.

The species are listed in the currently accepted taxonomic sequence published in the IOC World Bird List. See the note on bird taxonomy and bird names at the beginning of the book.

Few patterns provide any evidence with respect to taxonomy in these acuity data. For example, higher acuity is found in raptorial birds, but even among the Accipitridae (kites, hawks and eagles, Old World vultures) there is considerable interspecific variation, and the New World vultures (Cathartidae) have relatively modest acuity. It is also clear that there is considerable variation in acuity among passerine species.

The maximum spatial resolution recorded in the eyes of young humans (0.4 minutes of arc) is given at the foot of the table. It is clear that maximum human resolution exceeds that of the majority of bird species – but see Chapter 4 for discussion of the broader context of the acuity of birds relative to human acuity. In humans, acuity between 1 and 3 minutes of arc is regarded as 'mild vision loss', between 3 and 8 minutes of arc it is termed 'moderate visual impairment', and acuity lower than 10 minutes of arc is labelled 'severe visual impairment'. Thus, some of the species listed here would be regarded by human standards as having mild vision loss, while the owls have between mild loss and moderate impairment, and a Cormorant underwater would be regarded as having vision that is close to severe visual impairment.

Order	Family	Species	Acuity (minutes of arc)	Method
Struthioniformes	Struthionidae (ostriches)	Common Ostrich *Struthio camelus*	1.5	Ganglion
Tinamiformes	Tinamidae (Tinamous)	Chilean Tinamou *Nothoprocta perdicaria*	2.1	Ganglion
Galliformes	Phasianidae (pheasants and allies)	Grey Partridge *Perdix perdix*	2.9	Ganglion
		Japanese Quail *Coturnix japonica*	3.1	Ganglion
		Red Junglefowl *Gallus gallus*	3.6	Ganglion
		Common Pheasant *Phasianus colchicus*	2.3	Ganglion
		Indian Peafowl *Pavo cristatus*	1.5	Ganglion
Anseriformes	Anatidae (ducks, geese and swans)	Canada Goose *Branta canadensis*	3.1	Ganglion
		Canada Goose	1.8	Ganglion
		Northern Shoveler *Spatula clypeata*	2.7	Ganglion
		Gadwall *Mareca strepera*	3.0	Ganglion
		Mallard *Anas platyrhynchos*	2.5	Ganglion
		Greater Scaup *Aythya marila*	2.7	Ganglion
		Lesser Scaup *Aythya affinis*	2.7	Ganglion
		Red-breasted Merganser *Mergus serrator*	2.8	Ganglion
Columbiformes	Columbidae (pigeons, doves)	Rock Dove *Columba livia*	1.7	Behaviour
		Mourning Dove *Zenaida macroura*	3.9	Ganglion

Order	Family	Species	Acuity (minutes of arc)	Method
Procellariiformes	Hydrobatidae (northern storm petrels)	Leach's Storm Petrel *Oceanodroma leucorhoa*	4.2	Ganglion
	Procellariidae (petrels, shearwaters)	Northern Fulmar *Fulmarus glacialis*	0.7	Ganglion
Suliformes	Phalacrocoracidae (cormorants, shags)	Great Cormorant *Phalacrocorax carbo*	9.1	Behaviour (underwater)
Accipitriformes	Cathartidae (New World vultures)	Turkey Vulture *Cathartes aura*	1.9	Ganglion
		Black Vulture *Coragyps atratus*	1.9	Ganglion
	Accipitridae (kites, hawks and eagles)	Egyptian Vulture *Neophron percnopterus*	0.2	Behaviour
		Indian Vulture *Gyps indicus*	0.2	Behaviour
		Griffon Vulture *Gyps fulvus*	0.3	Behaviour
		Wedge-tailed Eagle *Aquila audax*	0.2	Behaviour
		Black Kite *Milvus migrans*	0.8	Behaviour
		Harris's Hawk *Parabuteo unicinctus*	1.0	Behaviour
Strigiformes	Tytonidae (barn owls)	Western Barn Owl *Tyto alba*	7.5	Behaviour
	Strigidae (owls)	Great Horned Owl *Bubo virginianus*	4.0	Behaviour
		Tawny Owl *Strix aluco*	2.7	Behaviour
Coraciiformes	Alcedinidae (kingfishers)	Laughing Kookaburra *Dacelo novaeguineae*	0.7	Ganglion
		Sacred Kingfisher *Todiramphus sanctus*	1.2	Ganglion
Falconiformes	Falconidae (caracaras, falcons)	American Kestrel *Falco sparverius*	0.75	Behaviour
		Brown Falcon *Falco berigora*	0.4	Behaviour
Psittaciformes	Psittaculidae (Old World parrots)	Bourke's Parrot *Neopsephotus bourkii*	3.2	Behaviour
		Budgerigar *Melopsittacus undulatus*	2.6	Behaviour

Order	Family	Species	Acuity (minutes of arc)	Method
Passeriformes	Meliphagidae (honeyeaters)	Brown Honeyeater *Lichmera indistincta*	1.5	Photoreceptor
		Red Wattlebird *Anthochaera carunculata*	0.7	Photoreceptor
	Acanthizidae (Australasian warblers)	Yellow-rumped Thornbill *Acanthiza chrysorrhoa*	1.2	Photoreceptor
	Corvidae (crows, jays)	Blue Jay *Cyanocitta cristata*	1.6	Ganglion
		Eurasian Magpie *Pica pica*	0.9	Behaviour
		Rook *Corvus frugilegus*	1.0	Behaviour
	Paridae (tits, chickadees)	Tufted Titmouse *Baeolophus bicolor*	4.5	Ganglion
		Carolina Chickadee *Poecile carolinensis*	0.6	Ganglion
	Zosteropidae (white-eyes)	Silvereye *Zosterops lateralis*	1.6	Photoreceptor
	Turdidae (thrushes)	Common Blackbird *Turdus merula*	1.3	Behaviour
	Muscicapidae (chats, Old World flycatchers)	European Robin *Erithacus rubecula*	5.0	Behaviour
	Passeridae (Old World sparrows)	House Sparrow *Passer domesticus*	6.3	Ganglion
	Fringillidae (finches)	Common Chaffinch *Fringilla coelebs*	1.3	Behaviour
		House Finch *Haemorhous mexicanus*	6.4	Ganglion
	Emberizidae (buntings, New World sparrows)	Yellowhammer *Emberiza citrinella*	3.1	Behaviour
		Common Reed Bunting *Emberiza schoeniclus*	3.8	Behaviour
Primates	Hominidae	Human *Homo sapiens*	0.4	Behaviour

Further reading

Published research papers that support all of the facts and ideas in this book are scattered across a very wide range of scientific journals. Many will be inaccessible to most readers for they are locked away behind the real and electronic access walls of academic libraries and academic publishers. However, if you want to read more and delve into more specialist literature, a good place to start is an online search of Google. Simply feed in species names and topics and refine your search to drill down to the most relevant published material. It will not throw up all relevant papers and the emphasis is upon more recently published work. However, such a search may throw up papers to which there is public access, and which can be downloaded or read on screen. Similarly searching topics or species on Wikipedia should throw up references that might be accessible through the internet or through public libraries.

Two books (both available in paperback) that will certainly lead the reader into a more detailed consideration of many of the topics discussed here are:

- Land, M. F. and Nilsson, D.-E. *Animal Eyes*, 2nd edition. Oxford University Press, 2012.

- Martin, G. R., *The Sensory Ecology of Birds*. Oxford University Press, 2017.

Two further books that discuss more generally topics within sensory ecology and consider a wide range of non-avian species are:

- Cronin, T. W., Johnsen S., Marshall, J. and Warrant, E. J. *Visual Ecology*. Princeton University Press, 2014.

- Stevens, M. *Sensory Ecology, Behaviour, and Evolution*. Oxford University Press, 2013.

Index

absolute visual sensitivity   79
   and night-time activity   177
Acrocephalus scirpaceus. See Eurasian Reed Warbler
acuity
   acuity-luminance function   29
   and colour   81
   comparing between species   76, 77, 259
   human   77
   and light level   29, 78
   and ultraviolet light   83
   See also spatial resolution
Aerodramus maximus. See Black-nest Swiftlet
Aerodramus spodiopygius. See White-rumped Swiftlet
Aethia cristatella. See Crested Auklet
Agelaius phoeniceus. See Red-winged Blackbird
Alopochen aegyptiaca. See Egyptian Goose
Amandava subflava. See Orange-breasted Waxbill
American Robin *Turdus migratorius*, lateral vision and
      foraging   96
amphibious vision   214
Anas acuta. See Pintail
Anas platyrhynchos. See Mallard
Anser caerulescens. See Snow Goose
Antarctic Prion *Pachyptila desolata*, olfaction and
      foraging   141
Apteryx australis. See Southern Brown Kiwi
Apteryx mantelli. See North Island Brown Kiwi
aquatic foraging   213
   and diving   218
   for mobile prey   222
   for sessile prey   220
   at the water surface   235
   through a water surface   218, 234
Aquila chrysaetos. See Golden Eagle
Ardenna pacifica. See Wedge-tailed Shearwater
Ardeola ralloides. See Squacco Heron
Ardeotis kori. See Kori Bustard
Atlantic Canary *Serinus canaria*
   audiogram   23
   sound location   125
Atlantic Puffin *Fratercula arctica*
   eye position   231
   foraging   233
   nostrils   137
   visual field   232
audiograms   22
   birds   121
   canary   23
   human   121

Aythya marila. See Greater Scaup

Bar-tailed Godwit *Limosa lapponica*, occasional
      nocturnal activity   208
Barn Owl *Tyto alba*
   contrast sensitivity   82
   ears   111
   nocturnality   182
   sensory ecology of foraging   19
   sound location   125
   visual field and outer ears   111
bill-tip organs   152
   in ducks and geese   153, 223
   in ibises   154, 156
   in kiwi   154, 156
   in parrots   155
   in shorebirds   155
binocular field
   in Eurasian Woodcock   8
   and lunging   108
   in Mallards   88
   in New Caledonian Crow   89
   in nocturnal birds   111
   and outer ears in owls   111
   and pecking   108
   in predators   111
   and seeing between the bill   108
   and seeing bill tip   108
   and tool use   109
   vertical extent   109
binocular vision
   function   107
   and optic flow-field   107
bird (definition), a bill guided by an eye   240
birds and berries   55
bird's-eye view   4
Black-Crowned Night Heron *Nycticorax nycticorax*,
      visual field and nocturnal activity   111
Black-nest Swiftlet *Aerodramus maximus*, sonar   133
Black Skimmer *Rynchops niger*
   bill breakage when foraging   237
   foraging and tactile cues   237
   foraging at the water surface   236
   occasional nocturnal activity   184
   visual field   111
Black-tailed Godwit *Limosa limosa*, vision and
      foraging   92

Black Vulture *Coragyps atratus*
    ears   127
    nostrils   137
Blacksmith Lapwing *Vanellus armatus*
    foot trembling   238
    visual field   238
Blue Crane *Grus paradisea*, collision vulnerability
    247
Blue Ducks *Hymenolaimus malacorhynchos*
    bill tip organ   223
    foraging   223
    tactile cues and foraging   223
    vision and foraging   101
Bobolink *Dolichonyx oryzivorus*, magnetoreception
    169
body odour. *See* semiochemicals
Brant Goose *Branta bernicla*, occasional nocturnal
    activity   208
Branta bernicla. *See* Brant Goose
Branta canadensis. *See* Canada Goose
Brown-headed Cowbird *Molothrus ater*, collisions
    252
Bubo virginianus. *See* Great Horned Owl
Bubulcus ibis. *See* Western Cattle Egret
Budgerigar *Melopsittacus undulatus*
    contrast sensitivity   82
    ears   127
    ganglion cell distribution pattern   73
    nostrils   137
    olfaction and species recognition   142
    sound location   125
    training, acuity   35
Buff-necked Ibis *Theristicus caudatus*, bill-tip organ
    157
Burhinus oedicnemus. *See* Eurasian Stone-curlew
bycatch. *See* collision hazards

Calidris alba. *See* Sanderling
Calidris alpina. *See* Dunlin
Calidris canutus. *See* Red Knot
Calidris maritima. *See* Purple Sandpiper
camera eyes
    arrangement in skull   49
    basic structure   46
    evolution   41
    functional components   46
    image analysis system   46, 51
    image properties   48
    optical system   46, 48
    variation   45, 47
Canada Goose *Branta canadensis*
    acuity   77
    linear area   74
Caprimulgus europaeus. *See* European Nightjar
Cathartes aura. *See* Turkey Vulture
Ceryle rudis. *See* Pied Kingfisher
Chicken *Gallus gallus domesticus*
    magnetoreception   168
    odour individual signature   143
    odour and reproduction   145
    tastes   161–3
Ciconia ciconia. *See* White Stork

Circaetus gallicus. *See* Short-toed Snake Eagle
clouds, effect on light levels   30
Cockatiel *Nymphicus hollandicus*, tastes, sweet   161
collision hazards   7, 246
    gill nets   247, 253
    glass panes   256
    power lines   247
    predisposing factors   249
    vehicles   247, 252
    wind turbines   247
collision mitigation
    diverting and distracting   251
    net lights   254
    warning markers   250
    window markers   257
colour vision
    colour through birds' eyes   59
    generality across birds   58
    and light level   44
    measuring discrimination   38
    'the rays … are not coloured'   44
    and spatial resolution   44
Columba livia. *See* Rock Dove
Columba palumbus. *See* Wood Pigeon
Common Chaffinch *Fringilla coelebs*, sound location
    129
Common Guillemot (Common Murre) *Uria aalge*
    dive depth   230
    diving at night   231
    diving for evasive prey at depth   220
    occasional nocturnal activity   184
    random searching for prey   233
    visual field   232
Common Kestrel *Falco tinnunculus*, eye structure
    18
Common Ostrich *Struthio camelus*
    eye lashes   106
    focal length   66
    optics   66
    sunshades   106
Common Poorwill *Phalaenoptilus nuttallii*
    nocturnality   204
    tapetum   205
Common Potoo *Nyctibius griseus*
    nocturnal activity   182, 206
    prey capture   206
Common Redshank *Tringa totanus*, occasional
    nocturnal activity   208
Common Shelduck *Tadorna tadorna*, occasional
    nocturnal activity   208
Common Snipe *Gallinago gallinago*, sound
    production   119
Common Starling *Sturnus vulgaris*
    eye focal length   66
    olfaction and selection of nest material   145
    optics   66
    sensory ecology of foraging   19
    sound location   124
    taste buds   159
    tastes   161–3
    visual field   67
compass. *See* magnetoreception

cones. *See* photoreceptors
contrast sensitivity
 birds 82
 cat 82
 human 82
control of bill position 93
Coragyps atratus. See Black Vulture
Corvus moneduloides. See New Caledonian Crow
Crested Auklet *Aethia cristatella*, olfaction and health
  detection 144

daytime, defined 30
Dendrocopos major. See Great Spotted Woodpecker
dimethyl sulphide (DMS), and foraging locations
  140
dioptre, defined 66
Dolichonyx oryzivorus. See Bobolink
Double-crested Cormorant *Phalacrocorax auritus*,
  occasional nocturnal activity 184
Dunlin *Calidris alpina*, taste and foraging 164

Eastern Rosella *Platycercus eximius*, olfactory bulb
  138
echolocation. *See* sonar
Egretta garzetta. See Little Egret
Egretta gularis. See Western Reef Heron
Egyptian Goose *Alopochen aegyptiaca*, occasional
  nocturnal activity 208
Epicurus 14
Erithacus rubecula. See European Robin
Eurasian Blackcap *Sylvia atricapilla*
 magnetoreception 210
 nocturnal migration 210
Eurasian Oystercatcher *Haematopus ostralegus*,
  occasional nocturnal activity 184, 208
Eurasian Reed Warbler *Acrocephalus scirpaceus*, song
  production 119
Eurasian Stone-curlew *Burhinus oedicnemus*, vision
  and foraging 108
Eurasian Wigeon *Mareca penelope*, vision and
  foraging 100
Eurasian Woodcock *Scolopax rusticola*
 bill-tip organ 155
 olfactory bulb 138
 panoramic vision 99
 remote touch 155
 vision and foraging 98
 visual field 88
 visual field and nocturnal activity 111
European Golden Plover *Pluvialis apricaria*, auditory
  foraging 209
European Nightjar *Caprimulgus europaeus*
 large gape 203
 nocturnal activity 182, 202
 rictal bristles 203
 touch sensitive palate 204
European Robin *Erithacus rubecula*
 fruits eaten 55
 magnetoreception 169
European Shags *Phalacrocorax aristotelis*, cone
  receptor pattern 71
eye lashes 106

eyes
 divergence of optic axes 88
 evolution 41
 position in skull 65
 size and imaging the sun 105
 what eyes do 43
 See also camera eyes

f-number 80
 and nocturnality 177
Falco peregrinus. See Peregrine Falcon
Falco tinnunculus. See Common Kestrel
Feral Pigeon. *See* Rock Dove
flicker fusion, measuring threshold 38
foraging
 underwater 222
 in water 213
fovea 72
Fratercula arctica. See Atlantic Puffin
Fringilla coelebs. See Common Chaffinch
Fulmarus glacialis. See Northern Fulmar
fundus oculi 68

Gallus gallus domesticus. See Chicken
ganglion cells
 distribution patterns 71, 74
 isodensity maps 52, 73
Garden Warbler *Sylvia borin*, magnetoreception
  169
Gavia immer. See Great Northern Diver (Common
  Loon)
geomagnetic field 166, 169
Geronticus eremita. See Northern Bald Ibis
Giant Kingfisher *Megaceryle maxima*, plunge diving
  for evasive prey 220
glass panes. *See* collision hazards
Glossy Ibis *Plegadis falcinellus*, bill-tip organ 157
Golden Eagle *Aquila chrysaetos*
 acuity 4
 eye structure 18
 vision and prey capture 90, 97
 world view 9
Golden Plover *Pluvialis apricaria*, visual field and
  nocturnal activity 111
Golden-winged Warbler *Vermivora chrysoptera*,
  infrasound detection 122
Grandry corpuscles 152
gratings and acuity measurement 34
Great Cormorant *Phalacrocorax carbo*
 acuity under water 77
 binocular vision between the bill 227
 diving for hidden prey 220, 224
 flush foraging 226
 foraging behaviour 224
 limits of information 8
 nostrils 137
 simulated vision under water 225
 training 34
 visual field 227
Great Grey Owl *Strix nebulosa*, limits of information
  8
Great Horned Owl *Bubo virginianus*, acuity 77

Great Northern Diver (Common Loon) *Gavia immer*, diving for evasive prey at shallow depths 220
Great Spotted Woodpecker *Dendrocopos major*, sound production 119
Great Tit *Parus major*
    caterpillar detection 1, 142
    olfaction and foraging 141
    olfaction and selection of nest material 145
    sound location 124
Greater Flamingo *Phoenicopterus roseus*
    feeding young 104
    filter feeding 104
    linear area 74
    visual field 104
Greater Prairie Chicken *Tympanuchus cupido*, sound production 119
Greater Scaup *Aythya marila*, diving to take sessile prey 220
Grey Plover *Pluvialis squatarola*, occasional nocturnal activity 208
Griffon Vulture *Gyps fulvus*
    collision vulnerability 246
    eye lashes 106
    eyebrow ridges 106
    visual fields 246
Grus paradisea. See Blue Crane
Gyps fulvus. See Griffon Vulture

Haematopus ostralegus. See Eurasian Oystercatcher
Haliaeetus albicilla. See White-tailed Eagle
Harris's Hawk *Parabuteo unicinctus*
    contrast sensitivity 82
    fovea 72
    sensory thresholds 37
    visual discrimination task 34
hearing 116
    dynamic range 120
    frequency range 120, 122
    kiwi 195
    threshold 119
Helmeted Guineafowl *Numida meleagris*, infrasound detection 122
Herbst corpuscles 151
Homing Pigeon. *See* Rock Dove
House Sparrow *Passer domesticus*, acuity 77
Humboldt Penguin *Spheniscus humboldti*
    diving for evasive prey at depth 220
    optics 228
hummingbirds, taste 161

image analysis 51
imaging the sun 91
Indian Peafowl *Pavo cristatus*, visual pigments 58
information
    and behaviour 6
    cost of information 11
    from sounds 116
    limits of information 8
    night driving 244
    paucity and cognition 243
infrared 26

infrasound 26
    Rock Dove detection 122

Kākāpō *Strigops habroptila*
    diet 200
    eyes 201
    nocturnal activity 182, 200
kestrel. *See* Common Kestrel
King Penguin *Aptenodytes patagonicus*
    dark adaptation 230
    dive depth 229
    diving at night 230
    dynamic pupil 229
    pin hole pupil 229
    visual field and nocturnal activity 111
Kiwi
    eye regression 197
    nocturnality 194
    sensory world 196
    See also North Island Brown Kiwi
Kori Bustard *Ardeotis kori*
    collision vulnerability 247
    visual field 86

Leach's Storm Petrel *Oceanodroma leucorhoa*
    linear area 74
    nostrils 137
    sensory ecology of foraging 19
Lesser Flamingo *Phoeniconaias minor*
    feeding young (crop milk) 104
    filter feeding 104
    visual field 104
light absorption by water 216
Limosa lapponica. See Bar-tailed Godwit
Limosa limosa. See Black-tailed Godwit
linear area 74
Little Egret *Egretta garzetta*, correction for water surface refraction 235
locating sounds 123–30
    direction 123
    distance (ranging) 124, 128
    sonar 131
Long-eared Owl *Asio otus*, outer ears 126
Ludwig's Bustard *Neotis ludwigii*, collision vulnerability 247

magnetoreception 166
    compass mechanisms 168
    magnetite model 169
    radical pair model 169
Malacorhynchus membranaceus. See Pink-eared Duck
Mallard *Anas platyrhynchos*
    bill tip organ 153
    taste buds 159
    visual field 88
Manx Shearwater *Puffinus puffinus*
    cone receptor pattern 71
    eye focal length 62
    eye structure, optics 62
    linear area 74
    nest location at night 211
    occasional nocturnal activity 184

photoreceptor density map   52
visual field   63
visual field and nocturnal activity   111
Mareca penelope. See Eurasian Wigeon
measuring senses   32
mechanoreception   150
Melopsittacus undulatus. See Budgerigar
migration, nocturnal   210
Molothrus ater. See Brown-headed Cowbird
moonlight   30
Morus bassanus. See Northern Gannet

natural illumination sources and light levels   30
Neotis ludwigii. See Ludwig's Bustard
Nestor meridionalis. See New Zealand Kākā
New Caledonian Crow *Corvus moneduloides*
    binocular field   89
    tool control   109
New Zealand Kākā *Nestor meridionalis*, vision and
        foraging   92
night-time
    challenges of nocturnality   174–8
    defined   30
    light levels   30, 176, 180
nightjars
    gape   203
    and nocturnality   202
    tapetum   204
    touch sensitive palate   204
    visual fields   204
nociceptors   152
nocturnality   172, 183
    kiwi   194
    and knowledge   188
    migration   210
        illuminated structures   210
    occasional   184, 208
    occasional nocturnal foraging   208
        sandpipers   208
        shorebirds   208
        waterfowl   208
    Oilbird   190
    owls   185
North Island Brown Kiwi *Apteryx mantelli*
    eye size   194
    hearing   195
    nocturnality   194
    olfaction   138, 140, 195
    touch   195
    visual field   198
Northern Bald Ibis *Geronticus eremita*, vision and
        foraging   103
Northern Fulmar *Fulmarus glacialis*
    ganglion cell distribution pattern   73
    linear area   74
Northern Gannet *Morus bassanus*, cone receptor
        pattern   71
Northern Lapwing *Vanellus vanellus*, auditory
        foraging   209
Northern Shoveler *Spatula clypeata*
    occasional nocturnal activity   184

panoramic vision   96
visual field   100
nostrils   137
Numida meleagris. See Helmeted Guineafowl
Nyctibius griseus. See Common Potoo
Nycticorax nycticorax. See Black-Crowned Night
        Heron
Nyctidromus albicollis. See Paraque
Nymphicus hollandicus. See Cockatiel

Oceanodroma leucorhoa. See Leach's Storm Petrel
oil droplets. *See* photoreceptors
Oilbird *Steatornis caripensis*
    cave dwelling   1
    eye structure   18
    f-number   191
    hearing   192
    nocturnal activity   190
    olfaction   193
    sonar   132, 192
    vision   191
    visual field and nocturnal activity   111
olfaction   135
    kiwi   139, 195
    location of foraging locations   140
    location of items   139
    navigation   146
    nest material selection   145
    nest recognition   145
    New World Vultures   139
    Oilbird   193
    petrels   139
    recognition of body condition   143
    recognition of individuals   143
    recognition of sex   142
    recognition of species   142
olfactory bulbs   136, 138
olfactory system   136
optic flow-field   94
optical adnexa   106
optical problems of immersion   214
Orange-breasted Waxbill *Amandava subflava*, eye
        structure, general   18
Ostrich. *See* Common Ostrich
owls
    knowledge and nocturnality   188
    and nocturnality   185
    and silent flight   186

Pachyptila desolata. See Antarctic Prion
panoramic vision   85, 96, 99
Parabuteo unicinctus. See Harris's Hawk
Paradise Shelduck *Tadorna variegata*, olfactory bulb
        138
Paraque *Nyctidromus albicollis*
    tapetum   205
    visual field and nocturnal activity   111
Passer domesticus. See House Sparrow
Pavo cristatus. See Indian Peafowl
pecking   101
Peregrine Falcon *Falco peregrinus*
    fovea   72

lateral vision and prey detection   113
   prey approach and capture   1
Phalacrocorax auritus. *See* Double-crested Cormorant
Phalacrocorax carbo. *See* Great Cormorant
Phalaenoptilus nuttallii. *See* Common Poorwill
Phoeniconaias minor. *See* Lesser Flamingo
Phoenicopterus roseus. *See* Greater Flamingo
photoreceptors
   cone abundance differences   71
   cone distribution patterns   70
   cone types   56, 59
   density   51
   isodensity contour maps   51, 52
   oil droplets   58
   rods   57
   sensitivity in the spectrum   57
   types   53, 57
Pied Kingfisher *Ceryle rudis*, correction for water
      surface refraction   235
pigeon. *See* Rock Dove
Pink-eared Duck *Malacorhynchus membranaceus*
   eye position in skull   50
   vision and foraging   101
   visual field   50
Pintail *Anas acuta*, panoramic vision   96
plastic ingestion   8
Platycercus eximius. *See* Eastern Rosella
Plegadis falcinellus. *See* Glossy Ibis
Plegadis ridgwayi. *See* Puna Ibis
Pluvialis apricaria. *See* European Golden Plover
Pluvialis squatarola. *See* Grey Plover
Poicephalus senegalus. *See* Senegal Parrot
Procellaria aequinoctialis. *See* White-chinned Petrel
psychophysics   32
Puffinus puffinus. *See* Manx Shearwater
Puna Ibis *Plegadis ridgwayi*, vision and foraging
      103
Purple Sandpiper *Calidris maritima*, taste and
      foraging   164

Red Knot *Calidris canutus*
   bill probing   1
   foraging techniques   102
   taste and foraging   164
   visual field and nocturnal activity   111
Red-winged Blackbird *Agelaius phoeniceus*, taste
      161–3
refractive power   66
relative sensitivity within senses   21
remote touch   155
resolution and light levels   29
resolution versus sensitivity   28
retina
   image analysis system   51
   photoreceptors   51
   tapetum   204
Rock Dove *Columba livia*
   acuity   77
   cone distribution pattern in retina   70
   contrast sensitivity   82
   eye focal length   62
   eye structure, general   18

eye structure, optics   62
f-number   80, 178
ganglion cell distribution pattern   73
homing   146, 167
infrasound detection   122
magnetoreception and navigation   167
odours and navigation   146
olfactory bulb   138
photoreceptor density map   52
spectral sensitivity   25
visual field   63
rods. *See* photoreceptors
Rynchops niger. *See* Black Skimmer

Sanderling *Calidris alba*
   sensory ecology of foraging   19
   taste and foraging   164
Scolopax rusticola. *See* Eurasian Woodcock
semiochemicals, species recognition   142
Senegal Parrot *Poicephalus senegalus*, bill-tip organ
      155
senses
   complementarity   242
   differences between species   16, 241
   dimensions   16
   intimate   148
   investigation   13
   telereceptive   115
   trade-offs between   242, 245
sensitivity versus resolution   28
sensory ecology   4, 6
sensory thresholds   20
Serinus canaria. *See* Atlantic Canary
Sextus Empiricus   14
Short-toed Snake Eagle *Circaetus gallicus*, visual
      sensitivity and resolution   28
Silvereye *Zosterops lateralis*, magnetoreception   169
smell. *See* olfaction
Snow Goose *Anser caerulescens*, linear area   74
somatic sensitivity   149–52
sonar   131
   Black-nest Swiftlet   133
   cave swiftlets   132
   human   131
   mammals   132
   Oilbird   132
   White-rumped Swiftlets   134
sound ranging   128
sounds
   frequency   120
   location   123–31
   physical characteristics   119
   produced by birds   117
Southern Brown Kiwi *Apteryx australis*
   bill-tip organ   156
   foraging   195
Southern Ground Hornbill *Bucorvus leadbeateri*
   eye lashes   106
   sun shades   106
spatial resolution, function   75
spatial vision, evolution   40
Spatula clypeata. *See* Northern Shoveler

spectral sensitivity   22
Spheniscus humboldti. See Humboldt Penguin
Spotless Starling *Sturnus unicolor,* olfaction, species
    and sex recognition   142
Squacco Heron *Ardeola ralloides,* correction for water
    surface refraction   235
starlight   30
starling. *See* Common Starling
Steatornis caripensis. See Oilbird
stereoscopic depth perception   93
Strix aluco. See Tawny Owl
Strix nebulosa. See Great Grey Owl
Struthio camelus. See Common Ostrich
Sturnus unicolor. See Spotless Starling
Sturnus vulgaris. See Common Starling
sun shades   91, 106
Super-senses   3
Sylvia atricapilla. See Eurasian Blackcap
Sylvia borin. See Garden Warbler

Tadorna tadorna. See Common Shelduck
Tadorna variegata. See Paradise Shelduck
Taeniopygia guttata. See Zebra Finch
tapetum   204
taste   158
    buds   159
    and foraging   164
    genes   160
    receptors   160
tastes
    bitter   163
    calcium   163
    fat   164
    salt   162
    sour   163
    sweet   161
    umami (amino acid)   161
Tawny Frogmouth *Podargus strigoides*
    nocturnal activity   182, 206
    prey capture   206
Tawny Owl *Strix aluco*
    eye f-number   80
    eye focal length   66
    eye position in skull   50
    nocturnal woodland environment   175
    nocturnality   178
    optics   66
    territoriality   189
    visual field   50, 67
    visual field and outer ears   111
    visual sensitivity and resolution   28
Theristicus caudatus. See Buff-necked Ibis
thermosensitive receptors   152
thresholds, measurement   36
touch   149. *See also* somatic sensitivity
trade-offs between senses   26
trade-offs within a sense   26
training birds   34
Tringa totanus. See Common Redshank
Turdus migratorius. See American Robin
Turkey Vulture *Cathartes aura,* olfaction   140
twilight, defined   30

two-choice discrimination   33, 34
Tympanuchus cupido. See Greater Prairie Chicken
Tyto alba. See Barn Owl

ultrasounds   26
ultraviolet   26
    and acuity   83
Uria aalge. See Common Guillemot

Vanellus armatus. See Blacksmith Lapwing
Vanellus vanellus. See Northern Lapwing
Vermivora chrysoptera. See Golden-winged Warbler
vision
    cost   197
    dynamic range   31
    evolution   40
    importance in birds   42
    regressive evolution   197
    sources of variation   45
visual fields   85
    and eye position in skull   50
    function   90
    human   86
    kiwi   198
    Kori Bustard   86
    and optic flow   94
    panoramic vision   85
    Western Reef Heron   89
    White Stork   86
visual illusions experienced by birds   12
visual sensitivity. *See* absolute visual sensitivity

Wedge-tailed Shearwater *Ardenna pacifica*
    acuity   77
    contrast sensitivity   82
    visual pigments   58
Western Cattle Egret *Bubulcus ibis,* non-correction
    for surface refraction   235
Western Reef Heron *Egretta gularis*
    seeing beneath the bill   89
    visual field   89
White-chinned Petrel *Procellaria aequinoctialis,* vision
    and foraging   92
White-rumped Swiftlet *Aerodramus spodiopygius,*
    sonar   134
White Stork *Ciconia ciconia*
    collision vulnerability   247
    sound production   119
    visual field   86
White-tailed Eagle *Haliaeetus albicilla,* vision   4
wind turbines. *See* collision hazards
Wood Pigeon *Columba palumbus*
    sound production   119
    vision and pecking   90
Woodcock. *See* Eurasian Woodcock
woodland canopy and light levels   178, 180

Zebra Finch *Taeniopygia guttata*
    olfaction and species recognition   142
    sound location   125
Zosterops lateralis. See Silvereye